URBAN CASCADIA AND THE PURSUIT OF ENVIRONMENTAL JUSTICE

URBAN CASCADIA AND THE PURSUIT OF ENVIRONMENTAL JUSTICE

EDITED BY

NIK JANOS AND CORINA MCKENDRY

UNIVERSITY OF WASHINGTON PRESS
Seattle

Urban Cascadia and the Pursuit of Environmental Justice was made possible in part by support from California State University, Chico, and Colorado College.

Design by Katrina Noble
Composed in Charis SIL, typeface designed by SIL International

25 24 23 22 21 5 4 3 2 1

Printed and bound in the United States of America

UNIVERSITY OF WASHINGTON PRESS
uwapress.uw.edu

LIBRARY OF CONGRESS CATALOGING-IN-PUBLICATION DATA
Names: Janos, Nik, editor. | McKendry, Corina, editor.
Title: Urban Cascadia and the pursuit of environmental justice / Nik Janos and Corina McKendry.
Description: Seattle : University of Washington Press, 2021. | Includes bibliographical references and index.
Identifiers: LCCN 2021004000 (print) | LCCN 2021004001 (ebook) | ISBN 9780295749365 (paperback) | ISBN 9780295749358 (hardcover) | ISBN 9780295749372 (ebook)
Subjects: LCSH: Environmental justice—Cascade Range. | Urbanization—Cascade Range. | Cascade Range.
Classification: LCC GE240.C36 U73 2021 (print) | LCC GE240.C36 (ebook) | DDC 304.209795—dc23
LC record available at https://lccn.loc.gov/2021004000
LC ebook record available at https://lccn.loc.gov/2021004001

The paper used in this publication is acid free and meets the minimum requirements of American National Standard for Information Sciences—Permanence of Paper for Printed Library Materials, ANSI Z39.48–1984. ∞

CONTENTS

ACKNOWLEDGMENTS

This book emerged out of many conversations between the editors about the contradictions and promises of green urbanism. We have both spent over a decade thinking and writing about this issue and were thrilled by the opportunity to bring together leading and emerging thinkers in this area and to do so with a focus on the region where these contradictions and promises are arguably most pronounced. Our greatest thanks go to the contributors to this project. We would also like to thank our institutions, California State University, Chico and Colorado College, for financial support of this work. Invaluable support has also been provided by Ben Pease at Pease Press Maps, Peter Hansen and the students at the California State University, Chico GeoPlace Mapping Lab, as well as Andrew Berzanskis, Neecole Bostick, and the staff of the University of Washington Press. We would also like to thank our families, who enable us to juggle the many hats of being teachers, scholars, spouses, and parents. We couldn't do this without you.

URBAN CASCADIA AND THE PURSUIT OF ENVIRONMENTAL JUSTICE

INTRODUCTION

Urban Cascadia and the Green Imaginary

NIK JANOS AND CORINA MCKENDRY

IF YOU HAVE EVER WATCHED THE SATIRICAL SHOW *PORTLANDIA*, you might recognize the mainstream imaginary of what some in the United States call the Pacific Northwest, or, expanding to include southern British Columbia, what we will be referring to in this book as Cascadia. In the opening sequence of the show, we get it all: green lens hue; urban landscapes nestled among dense forests, rivers, and looming mountains and a volcano; hipsters riding bikes; local bookstores; offbeat counterculture and REI yuppies; and western North American colonial architecture, symbolized by craftsman and bungalow homes and brick and steel industrial cityscapes. These scenes and the show's running skits bring to life a mostly white, middle-class, green urban romanticism yet evoke a sense that there is, contra Gertrude Stein, a "there there" and that Cascadia has a sense of place, culture, and history that is distinct from other parts of North America.

Cascadia holds an imaginary as the center of new forms and practices of green urbanism and environmental sustainability. This imaginary arises from its literal greenness situated in the vibrant biomes on either side of the Cascadia Mountains, as well as from the particular forms of green urbanism and environmental governance that shape the politics of the region. This book will undertake a critical examination of the real and imagined greenness of Cascadia by paying attention to the effects of what we consider the three core historical and contemporary developments at play in the region: capitalist urbanization of nature, stretching from European colonialism to contemporary green urbanism; the creation of social and environmental inequalities that contradict the green, progressive image of the region; and the movements and efforts to construct social and environmental justice out of the first two. Though these three core developments are not exclusive to Cascadia, the strength of the green imaginary in the region makes them particularly pronounced.

This book explores the ways these histories, inequalities, and relationships with more-than-human nature shape how the people of Cascadia navigate the political and economic barriers and contradictions they face in constructing a more socially and environmentally just region. We describe this region, taken together, as *urban Cascadia*. By this we mean not just the major cities of the region but the infrastructure, social networks, built and natural environments, and nonhuman animals and plants that are all interconnected in an increasingly urbanized bioregion, which is also intersected and shaped by national and planetary processes and institutions. Telling neither a story of green disillusion nor, its opposite, a story of green boosterism, this book grapples with the contradictions of urban Cascadia and how these contradictions help us understand broader questions of democracy, social justice, urban governance, and the politics of environmental change.

Some primary questions that guide this book are the following: If indeed there is a "there there," what do the unique historical and contemporary qualities relating to the social and ecological region people call Cascadia have to tell us about the processes and patterns of capitalist urbanization? In what ways is settler colonial society acknowledging (or failing to acknowledge) the realities of colonialism, and what steps are being taken by Indigenous and non-Indigenous people to construct genuine politics of recognition, reconciliation, and Indigenous sovereignty throughout the region? What ways are people in Cascadia addressing social and environmental inequities, environmental degradation, resource use, and climate change and decarbonization, and what are the barriers faced in achieving a more democratic, just, and sustainable region? What can in-depth research and thinking about Cascadia tell us about environmental and urban politics, environmental justice, the relationship between cities and rural areas, and the politics of climate change more broadly? In exploring these questions, the contributors to this book build a dialectical sense of place of Cascadia by applying a critical lens to its myths and imaginaries, to its structural inequalities and contradictions, and to its promise.

WHERE AND WHAT IS CASCADIA?

Despite *Portlandia* and other popular imaginaries of the region giving us a sense that there is something unique about this slice of North America, there is no singular Cascadia. Instead, various actors, including the bioregionalist Cascadia Institute, elected city and regional officials in both the United States and British Columbia, and public-private entities such as the Cascadia Innovation Corridor, construct Cascadia as a political ideology and project. However, Cascadia does not exist as an actual political territory or a coherent sociopolitical object. It is a place of becoming, and as Jeffrey Sanders shows in chapter 1, it is quite often used and imagined as a utopia. As an imaginary place, Cascadia is often an ideology, one that is, of course, contested, divergent, and unsettled.

In his 2004 article "Transborder Cascadia," Patrick Smith unpacks this definitional uncertainty by separating Cascadia into "physical, functional, and ideological" definitions.[1] A physical definition of Cascadia bases Cascadia's boundaries on geophysical and climatological characteristics, typically as the region from the Pacific Ocean on the west and the Rocky Mountains on the east, with the heart on either side of the Cascade Mountains from British Columbia down to the northernmost part of Northern California (see map I.1). The functional definition attempts to locate Cascadia in terms of the infrastructural, territorial, political, and economic roles it plays in an interconnected regional, national, and international space. Defined ideologically, Cascadia is the shared values, ideologies, and identities such as freedom, democracy, and environmentalism that are, this definition asserts, endemic to the biome called Cascadia.

Synthesizing Smith's illustration of the definitional uncertainty of the region, in this section, we will discuss two of the most prominent understandings of Cascadia, those that underlie the chapters of this book. We start with an exploration of Cascadia as an ideology of bioregionalism, which formed via an integration of the physical and social sciences. We then describe what we are calling eco-capitalist Cascadia, which embodies a merging of functional and ideological definitions.

THE ECO-POLITICS OF BIOREGIONAL CASCADIA

The ideological definition of Cascadia was created during the heyday of bioregionalism in the 1970s. Bioregionalism is a political

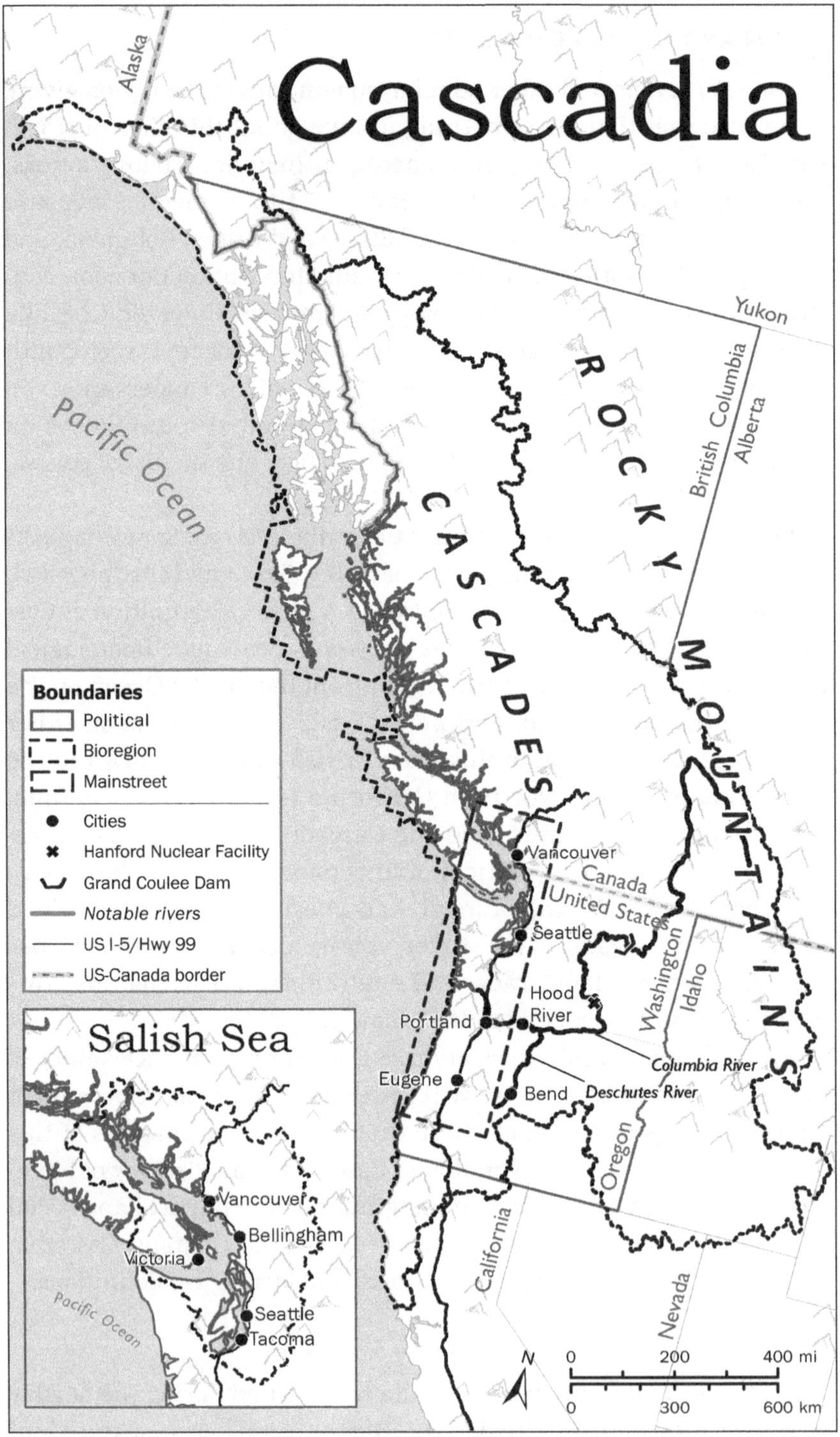

Cascadia
Alaska
Yukon
British Columbia
Alberta
Pacific Ocean
ROCKY MOUNTAINS
CASCADES
Boundaries
Political
Bioregion
Mainstreet
Cities
Hanford Nuclear Facility
Grand Coulee Dam
Notable rivers
US I-5/Hwy 99
US-Canada border
Vancouver
Canada
United States
Seattle
Washington
Idaho
Hood River
Portland
Columbia River
Eugene
Bend
Deschutes River
Oregon
California
Nevada
Salish Sea
Vancouver
Bellingham
Victoria
Seattle
Tacoma
Pacific Ocean
N
0
200
400 mi
0
300
600 km

philosophy and movement that critiques industrial capitalism's role in damaging and degrading the Earth's biosphere, including human and nonhumans.[2] Bioregionalism holds that to rectify these effects, political movements arise from the distinct bioregion in which they are situated. By defining a region by shared natural features rather than the artificial political boundaries made by politics, it is argued, there will be a greater inherent desire to establish economic and social systems that protect the local and regional ecology.

Intellectuals and environmentalists have long asserted that Cascadia is one such naturally existing bioregion. The term *Cascadia* is generally considered to have been coined by David McCloskey, a sociologist at Seattle University. McCloskey went on to create an educational nonprofit called the Cascadia Institute.[3] In advancing the political imaginary of Cascadia, the Cascadia Institute critiques other names and conceptions of the region as "a construction—a word or phrase fashioned by the powers that be (or would be), and imposed on the land to take control over it, to impose a 'territory' or 'ideology' or 'brand' on the place itself."[4] For McCloskey and others, bioregional Cascadia is a strategically biocentric, essentialist, and ecologically determined politics. The institute writes, "Cascadia is first encountered as a real place, more a living body than an image or idea. For Cascadia is a land rooted in the earth, and animated by the turnings of sea and sky, the whole mid-latitude wash of winds and waters. . . . The land we seek to understand is both a physical terrain and a terrain of consciousness."[5] To put it another way, in these conceptions of Cascadia, the politics flows from the ecology rather than the ecology being co-constructed by politics.[6]

Fifty years on, there are additional organizations dedicated to furthering the imaginary and politics of bioregional Cascadia. For example, *Cascadia Magazine* describes itself as "a nonprofit publication that is passionate about crossing borders between nations, cultures, ethnicities, and gender—as well as bridging the rural/urban divide."[7] They define Cascadia as "an effort to understand a larger bioregion that extends beyond the straight lines of national borders. Having a bioregional approach encourages thinking about the entire ecosystem (including the human population within it). . . . That's it. You live here: no nationality or citizenship matters. You're part of a

MAP I.1. Cascadia. Map by Alan M. Vigent and Peter Hansen.

local ecosystem."[8] CascadiaNow!, a nonprofit social movement organization, defines Cascadia as "a positive and inclusive, place-based movement focused on building autonomous and equitable local infrastructure that is both resilient and sustainable. The movement is based on the idea of transcending arbitrary state borders and shifting our actions and impacts locally."[9] CascadiaNow! argues that the Cascade movement's mission, rooted in bioregional and local political identities and networks, is to "further local autonomy, empower individuals and communities to better represent their own needs, and create sustainable local economies through bioregional planning."[10]

The bioregional mythology and ideology of Cascadia can be characterized as the merging of regionalist and localist ideology with environmentalism and ecology. As discussed above, in this merging, politics, space, and ecology are naturalized. Therefore, the political ethos of Cascadia—such as local democracy, autonomy, subnational sovereignty, mutual aid, freedom, social and natural harmony, inclusion, and mutual interests of naturalized inhabitants—is assumed to spring from the biome in which the inhabitants reside. The imagined community of Cascadia is *sui generis*, naturally determined, and stands in contrast to national and global relations, states, and ideologies.

Although the proponents of bioregionalist Cascadia purposely borrow from natural history and Indigenous history and ideas in their formulation, this ideology is in fact most prominent among white-collar urban environmentalists.[11] Coupled with conservationist and preservationist goals, as well as the middle-class idea of nature as a site of recreation, bioregional Cascadia makes ideological alliances with the other mainstream ideology of Cascadia, eco-capitalism.[12] Specifically, the meeting point between the two is the assumption of a naturally and ideologically coherent place called Cascadia.

THE SPATIAL POLITICAL ECONOMY OF ECO-CAPITALIST CASCADIA

The eco-capitalist Cascadia imaginary formed out of the changing political-economic practices and ideologies associated with the economic globalization of the past forty years.[13] Specifically, even as globalization constructed global linkages and institutions, it simultaneously engendered the growth of regional and local linkages and institutions.[14] The formalization of the North American Free Trade Agreement (NAFTA) in 1994 is an example of this rescaling, as

Canada, the United States, and Mexico reconfigured the power of the nation-state to regulate economic practice to a megaregional trade agreement between the three countries. On the border between Canada and the United States, NAFTA provided national cover to pursue smaller regional political-economic coordination between the provinces and subnational states of these two countries. The area emerged as a cross-national regional political-economic bloc with Cascadia as a reigning brand mythology.[15]

There are a number of ways to conceptually as well as politically draw eco-capitalist Cascadia. The first is what Smith calls "traditional Cascadia," which consists of British Columbia, Canada, and the US states Washington and Oregon.[16] Traditional Cascadia is an embodiment of the subnational imagination and territorial governance of a slice of the continent, beginning and ending at the subnational state/provincial borders. More germane to questions of urbanization is what is referred to as Mainstreet Cascadia. This definition merges the bioregionalist definition, discussed above, with the idea that an urbanized megaregion spans from Vancouver, BC, down the British Columbia Highway 99/US Interstate 5 corridor to Eugene, Oregon, but also interconnects the entire bioregion via transportation and energy infrastructure, economic ties, labor markets, commodity flows, culture and history, and governance agreements, among other things.

It is worth noting that twenty years ago, prolific scholar of the region Carl Abbott raised serious doubts about the institutional formation of a Cascadia megaregion. Barriers he saw included the strength of existing place, national, and economic attachments as well as the spatial distance between cities. However, despite his skepticism that it would do so, in intervening years the salience of eco-capitalist Cascadia has only become stronger.[17] In many ways the megaregion has become the ideological pinnacle of eco-capitalist Cascadia. It is an idealized space that merges the imaginary of an essential cultural and material nature with the political economy of urbanization throughout the region.

While elected officials and elites have used Cascadia as a political imaginary for decades,[18] in 2016, British Columbia premier Christy Clark and Washington governor Jay Inslee took eco-capitalist Cascadia branding to a new level. They created and signed a nonbinding but discursively clear memorandum of understanding (MOU) between British Columbia and Washington State that created the Cascadia Innovation Corridor (CIC), formally naming the geographic and political ideology of the region. According to Clark, the MOU "strengthen[s] the existing tech ties between British Columbia and Washington state [which] is a priority for both our governments. By working together, the economic opportunities, job growth and advancements in the sector will place us at the forefront of innovation."[19] They agreed to facilitate trade; encourage partnerships that focus on high-tech industries, including life sciences, clean technology, and data analytics; promote advanced research; endorse education programs to train a new high-tech workforce; and strengthen regional transportation planning.[20]

The CIC is a partnership between for-profit, nonprofit, and governmental organizations. Its steering committee is a who's who of elites in and out of the private sector, including former Washington governor Christine Gregoire, heads of business councils and chambers of commerce, and executives from major corporations, including Microsoft. In fact, Microsoft was one of the early proponents of the CIC. Since its founding in 2016, the CIC has expanded to include Portland. Mainstreet Cascadia, defined above, is moving closer to becoming a literal place with institutional vision and planning. The CIC's vision of Cascadia is thus: "Vancouver, B.C., Seattle and Portland have much in common: unsurpassed beauty; proximity to Asia; and market-leading capabilities in key economic sectors. By linking these cities, the Cascadia Innovation Corridor will create opportunity and prosperity beyond what they and their surrounding regions could achieve independently. Together, we can be the next global innovation ecosystem and a leader in solving some of the world's toughest challenges. Together, we are the Cascadia Innovation Corridor."[21] Proponents of eco-capitalist Cascadia are using the concept of Cascadia to brand the region and the cities within it. In the steady stream of websites, speeches, and press material since 2016, eco-capitalist Cascadia is defined as a set of political and economic linkages and policies to connect the region via high-tech corporations; improved transportation, such as the proposed high-speed rail project that would drastically collapse time and space between British Columbia, Seattle, and Central Oregon; and corporate research partnerships with public universities.[22]

Of course, COVID-19 temporarily derailed these dreams of a seamlessly integrated region, making it clear the extent to which the emergence of eco-capitalist Cascadia continues to rest on the politics of the nation-state. The ban on nonessential travel between Canada and the United States, implemented in March 2020 and still in place as of June 2021, dramatically decreased travel and associated economic activity across the region.[23] The differences in the federal responses to the pandemic, with Canada's more robust actions leading to notably better control of the virus, have also brought differences between the two countries into stark relief.[24] Yet the work toward an eco-capitalist Cascadia continues, with the preface to the CIC's new long-term vision document, released in September 2020, stating, "It is clear this disease has upended lives, but Cascadia is resilient. While we continue to fight COVID-19 and its consequences, we also need to address the undeniable fact that our mega-region is not prepared for the future to come. Protecting the health of our community and ending this pandemic are paramount, but we cannot deny that the challenges outlined in this report remain, and the responsibility rests with us to ensure Cascadia is ready to address them."[25] Indeed, the report asserts, the pandemic has made the importance of establishing a sustainable and resilient region even more pronounced.

As they enter the third decade of the twenty-first century, decision makers and regional elites are attempting to build the institutions, networks, and infrastructure that have the potential to shift Cascadia from a branding device to a new regional political scale connecting cities, provinces, and nations. The imaginary of eco-capitalist Cascadia therefore has material consequences as it is used to gain competitive advantage through a reterritorialized economic region that can secure the benefits of the high-tech, globalized economy. It is ultimately an elite imaginary used to further the interests and projects of local, regional, and global corporations and the interests of urban politicians and boosters.

This book takes as a starting point these conflicting and unsettled definitions and practices of left bioregional Cascadia and eco-capitalist Cascadia. They both contain a certain discursive power, as can be seen in the ongoing attempts to use them to create a real political, economic, cultural, and ecological object of the region. Yet the book then takes a critical approach to the intertwined physical, functional, and ideological implications of these definitions of Cascadia. The authors in this book challenge the boosterism of both of these approaches, in particular the use of Cascadia as a post-political erasure of power and inequality via the projection of a nominally democratic, but in practice elitist, naturalized green imaginary.[26] Instead, the authors strive to unpack the relationships between these imaginaries of a coherent, dynamic, progressive region with the history of environmental destruction, colonialism, racism, and inequalities that have shaped Cascadia since the earliest days of settler colonialism. By building politics out of a predetermined nature, bioregional and eco-capitalist Cascadia essentialize and naturalize politics. In contrast, this book follows the insights of urban political ecology and "explicitly recognizes that the material conditions that comprise [socionature] are controlled, manipulated and serve the interests of the elite at the expense of marginalized populations. These conditions, in turn, are not independent from social, political and economic processes and from cultural constructions of what constitutes the 'urban' or the 'natural.'"[27] The point, say political ecologists, is to politicize nature, not naturalize politics.

THE CONTRADICTIONS OF CASCADIA

URBANIZATION

Eco-capitalist Cascadia defines the region as an extended and hyperconnected urbanized megaregion. While proponents of this definition are largely focused on the interconnected cities on the Highway 99/I-5 corridor, this book seeks to understand the myriad forms of urbanization and environmental transformation of Cascadia that originate in yet also extend beyond the region's major cities. To do so, we draw on urban theorists Neil Brenner and Christian Schmid's theoretical framework of "planetary urbanization" or "extended urbanization."[28] Brenner and Schmid challenge those interested in cities, the urban, and urbanization to move beyond a definition of urbanization that is based on a typology of cities as mass human settlements and agglomerations of buildings. Rather, they argue that such definitions of the city should be replaced by an understanding of the *process* of capitalist urbanization that is reshaping spaces, territories, and places far outside the city, as well as the city itself. To this end, a number of the authors in this book consider how urbanization is stitching together rural and wilder spaces, processes, and people well outside the city (see chapters 2–4).

Exploring this thread even further, the chapter by Nik Janos (chapter 2) and our coauthored chapter (chapter 12) investigate how processes of planetary urbanization connect Cascadia into the circuits and processes of capitalist globalization. The power of the planetary urbanization thesis is in recognizing, for example, how the colonial and industrial resource extraction of the late nineteenth and early twentieth centuries in British Columbia, Washington, and Oregon, which consisted of trees and salmon as well as coal and other minerals, fueled industrial and urban development in faraway places. Or, to a more recent example, how the Cascadia-based Microsoft and Amazon corporations are major players in the globalized IT economy, reshaping everyday life, labor, and urban infrastructure across the planet (see chapters 1, 2, and 9).

Some have taken Brenner and Schmid to mean that studying the city is a dead end; however, that is not our interpretation. While the planetary urbanization thesis is a useful tool, this book will also explore the cities of Cascadia from the dimension of the politics of territory and sovereignty (chapter 12) as well as sites of multiscalar urban and environmental governance (chapters 7 and 11).[29] The city is a primary site of everyday life, and its significance therefore cannot be overlooked. Laypeople and scholars continue to have a fascination with and focus on the cities of Cascadia for the important cultural, political, and economic centrality they are assumed to have. Rather than misguided city-centrism, there are many reasons to continue to excavate the mythologies and the existing built environments of the cities of Cascadia. Yet as discussed in the opening paragraphs, our urban Cascadia framework also recognizes the many dimensions of capitalist urbanization of nature along the spatial continuum, from so-called wilderness to suburban/rural interface to cities.

RACE, CLASS, AND SETTLER COLONIALISM

Like the rest of North America, Cascadia is deeply scarred by the effects of settler colonialism. Chapters by Coll Thrush (chapter 5) and Danica Miller (chapter 6) underscore one of the primary contradictions of Cascadia's image as a bastion of progressive politics: the legacy of and continued dispossession of Indigenous land and rights of

MAP I.2. Selected Indigenous linguistic communities and cultures of the Salish Sea and Columbia River. Map by Ben Pease.

self-determination (see map I.2 for the traditional territories of the Indigenous people of the Salish Sea and lower Columbia River). As Miller notes, the struggles of Indigenous peoples, and their battles against colonizers and the US and Canadian governments, largely have taken place over land and resources, such as the Fish Wars on the Puyallup River. However, as Thrush tells us, the experience of Indigenous communities in Cascadia is not exclusively a rural one; in fact, it is significantly an urban story. This book, therefore, weaves two interrelated and equally important stories: the dispossession of land and resources and the myriad ways that Indigenous people and their allies have struggled and continue to struggle for self-determination and to contribute deeply to the environmental politics of the region (see chapter 7).

The condition of inequality of non-Indigenous communities of color is another structural aspect of Cascadia that defines the region.[30] Oregon was conceived specifically as a white supremacist state. The Oregon Constitution of 1857 prohibited Blacks from residing there and denied people of color equality before the law until well into the twentieth century.[31] Oregon was a hotbed of resurgence of the Ku Klux Klan after World War I and of Nazi and fascist groups in the 1930s and beyond.[32] As such, the white supremacist groups that took to the streets of Portland in response to Black Lives Matter protests in the summer and fall of 2020 were by no means an incongruity in a bastion of progressivism as they were sometimes portrayed.[33] Oregon is not alone in its appeal to white supremacists, however. In the decades since the bioregional idea of Cascadia was conceived, right-wingers, white supremacists, and ethno-regionalists in Oregon, Washington, Idaho, and Montana have adopted Cascadia as a proposed territory of white separatism.[34]

British Columbia too has a long history of racial exclusion. Almost all of British Columbia is unceded Indigenous territory. The province's racist history also includes a Chinese Exclusion Act, Japanese internment, laws banning certain ethnic groups from voting and immigrating, and the physical erasure of communities such as Black Vancouverites' Hogan's Alley (see chapter 8).[35] White supremacist groups are also active across Canada, though they are "more loosely organized and less focused on gun rights" than their US counterparts.[36] There are estimated to be over a hundred active white supremacist groups in Canada, and their activity has been increasing over the past several years. Furthermore, the BC human rights commissioner has noted numerous racially motivated hate crimes since the COVID-19 pandemic began.[37]

The Cascadia imaginary is a generative space for racist groups, particularly in the United States, because it provides two key ingredients that have historic roots in white supremacy, Nazism, and fascism: nature worship, including assigning to nature the values of purity and essence, and territorial separatism, defined here as the idealized space for the pure people of naturalized territory.[38] Coupled with the historical and contemporary demographic reality of the Cascadia region, in which the US parts of the region have a white supermajority, we can see that Cascadia is a vessel that divergent ideologies can pour themselves into. As Signe Marie Cold-Ravnkilde, Jaidev Singh, and Robert G. Lee observe, the Cascadia mythos fits well with the project of "othering," in which "competing geographical imaginings distinguish who belongs and who does not belong to the image of Cascadia."[39] Leftist environmentalists and eco-capitalists alike have to contend with the reality that Cascadia by its bioregional nature is a highly elastic concept and is open to appropriation from right-wing, racist, ethno-religious regionalist and nationalist groups.

INEQUALITIES

As the histories of racism of the region challenge the progressive, green image of Cascadia, so too do historical and ongoing environmental injustices. Matthew Klingle documents how the geography of racial, ethnic, and class inequalities was built into the landscape, with working-class and poor Greeks and Italians, African Americans, Indigenous peoples, Swedes, Japanese, and Russian Jews squatting in the tidal areas along Seattle's Elliott Bay and the Duwamish River at the turn of the twentieth century.[40] A similar story is told by Robert McDonald in his chronicle of the multi-ethnic making of Vancouver and the ways that English elites built environmental inequalities into the fabric of the city. Rough living, coupled with rough working conditions, made life for the working class perilous, even as the urban elite poured vast amounts of capital into private property, factories, terraforming and moving hills, and building railroads to get lumber, coal, and fish to regional, national, and global markets.[41]

These scenes were common up and down the coast and into the interiors of Cascadia. Capitalists transformed the socioecologies of the region: leveling hills, filling tidal flats and estuaries, felling forests, and, perhaps most significantly, damming rivers, including turning one of the largest rivers in North America, the Columbia River, into what Richard White calls an "organic machine."[42] These transformations enabled the rapid harvest of salmon, regional electrification, urbanization, and the "greening" of the drier eastern side of the Cascade Mountains, turning it into the region's breadbasket. Timber and mining towns sprung up throughout the region, facilitating the movement of commodities from the hinterlands to the port cities. Dams and electrification projects further stitched hinterlands and cities into an assemblage of urban nature (see chapter 3). Today, nearly a hundred years since the heyday of the area's extractive economy, the legacy of these environmental and social inequalities is observable on the landscape as well as in the ethnic and racial composition, poverty and wealth, and the urban and rural political alignments of Cascadia (see chapter 1).

However, in the intervening century, the dynamics of environmental inequalities have changed. Today, there is a growing crisis of housing affordability, particularly in Vancouver and Seattle (see chapters 8–10), as well as substantial inequalities around jobs and wages, health, and environmental pollution. In the cities, elites have gone full press on creating technology-heavy economies that require a highly skilled workforce. High-tech sector wages coupled with a frenzy of urban redevelopment and the parking of global capital in the regional housing market have led to rising housing costs, making it hard for working-class people to afford to live in these cities. And, when they do, they are often restricted to heavily polluted but more affordable neighborhoods, such as Seattle's Georgetown and South Park, where residents "have up to a 13-year shorter life expectancy (at birth) than wealthier parts of Seattle."[43] According to the Children's Alliance, 15 percent of children in Washington live in poverty, with Black children three times and Indigenous children four times as likely to be in poverty as white children.[44] In British Columbia, racialized groups, particularly immigrants, are significantly more likely to be living in poverty than native-born white Canadians.[45]

There is also a widening gap in income, health, and opportunity across the urban to rural continuum of Cascadia. The gap between the residents in the core metro areas and those in more rural areas is largely between richer urban and suburban whites and poorer suburban to rural whites. In Oregon, for example, the largest concentrations of poverty are in small towns in rural areas, with notable gaps in income, education levels, child poverty, and adult suicide rates between what the state categorizes as urban versus rural and frontier counties.[46] Rural areas and resource-based small towns in British Columbia likewise struggle with poverty and intersecting deprivations.[47] All told, twenty-first-century inequalities span multiple geographic and political territories throughout Cascadia, affecting Blacks, Latinx people, immigrants, and Indigenous people in the central cities and suburbs as well as Indigenous people and whites in the suburbs, exurbs, small towns, and rural areas. Such inequalities sit uncomfortably with the progressive politics of the region.

THE PROMISE OF CASCADIA

As in most places on Earth in the so-called Anthropocene, capitalism, urbanization, and industrialization have radically transformed Cascadia and have given rise to deep-seated social and environmental injustices. However, despite the frequent mismatch between the idea of green Cascadia and the structural conditions of environmental degradation and injustice, the people of the region have also been notable exemplars of progressive environmental thought and action in the late twentieth and early twenty-first centuries. These include a wave of protests and fish-ins in the 1960s to 1970s that elevated tribes to major stakeholders in environmental and salmon management (see chapter 6); the creation of the Washington Department of Ecology in 1970 and the passage of the Growth Management Act in 1990, both of which led to new scales of environmental governance in Washington; the 1999 endangered species listing of salmon, making Seattle and Portland the first cities to have such a listing; the birth of radical environmental organizations like Greenpeace, founded in Vancouver in 1971; the forest and ecosystem direct actions in the 1980s and 90s across the region; the numerous local justice organizations that have worked to make urban environmental remediation synonymous with social and racial justice (see chapters 7, 9, and 11); major commitments to carbon neutrality alongside commitments to reconciliation with First Nations in Vancouver; and interventions by governments throughout the region aimed at stopping the expansion of fossil fuel infrastructure (see chapter 12). In sum, the people of Cascadia are often leaders in environmental discourse and action that have implications beyond the region. Therefore, this book will amplify these stories to parse out what we can learn from their successes and failures and to illuminate how they enable us to better understand environmental and urban governance, policy, law, and social participation by non-state actors.

GREEN URBANISM AND ENVIRONMENTAL GOVERNANCE

Vancouver, Seattle, and Portland are frequently celebrated in the popular press as leading innovators in low-carbon and green urbanism.[48] Such accolades are not unfounded. These cities have been global leaders in rethinking high-carbon, sprawling urban development models; protecting green space; promoting green transit; and requiring the private sector to take strong action to reduce carbon emissions from the built environment. As such, they are seen as offering hope for reducing carbon emissions in the face of inadequate federal action on climate change in both countries.

The popular attention given to the environmental agendas of these green cities has been accompanied by a larger scholarly inquiry into the role of subnational governments in addressing environmental issues, particularly climate change. This literature has expanded from an examination of what Harriet Bulkeley and Michele Betsill call "municipal voluntarism"—the efforts of city governments to lower the environmental impact of municipal activities—toward a broader understanding of local governments' environmental contributions. These include engagement with the private sector and other subnational governments as well as recognition by local officials of the need to participate in a multiscalar process of climate governance to achieve climate mitigation and adaptation goals.[49]

Taking this even further, there has been a growing interest in municipal governments as just one part of a complex social and technological system that also includes social movements and civil society, past infrastructure choices, globalized financial markets, and even the ecosystems in which urban places and processes are embedded.[50] This shift in understanding of local climate and environmental politics has opened up further analytical space for understanding the opportunities and challenges of transforming such systems toward ones of greater environmental sustainability and social justice. As will be seen in the chapters that follow, however, this is no easy feat.

ENVIRONMENTAL JUSTICE AND DEMOCRACY

Even as scholars of urban climate governance are increasingly seeing city efforts as part of broader sociotechnical processes of furthering environmental sustainability, other critiques have emerged about the localized impacts of cities' green development efforts. Especially in cities with tight housing markets in which gentrification pressure is already pronounced (including all the major cities of Cascadia), there is evidence that the environmental amenities that enhance the greenness of a city and reduce its carbon emissions, particularly parks and green transportation options, can exacerbate gentrification in neighborhoods closest to those amenities.[51] Other factors that define the socioeconomic context of Cascadia also contribute to gentrification and displacement (see chapters 7–10). These include a shortage of affordable housing, the boom in the tech industry and accompanying wage inequality, decades of reductions in federal and provincial housing assistance, and the price pressures created by global investment capital being pumped into housing markets. But on a very local level, environmental efforts are also contributing to displacement of low-income residents, and the destruction of long-established communities of color has raised vital questions about for whom the eco-capitalist green city is for.

This literature on green gentrification and climate gentrification is part of a larger expansion of how environmental justice is understood. The traditional focus of environmental justice activists and scholars on the disproportionate exposure of communities of color to pollution and environmental degradation certainly remains important (see chapters 7 and 11). Yet such analyses have also expanded to include justice in access to environmental "goods," from green space to quality affordable housing, *and* a demand that such goods serve communities that have been deprived of these vital amenities for decades, rather than these amenities furthering their displacement.[52] In addition, environmental justice activists and scholars insist that this distributive justice will not be achieved without robust, deeply democratic participatory processes in which communities play a fundamental role in defining both the problems of and the solutions to environmental injustice.[53]

Critical environmental justice scholars are also increasingly asserting the place of more-than-human nature in analyses of environmental (in)justice, arguing that "the harms suffered by ecosystems today are closely linked to and mirror the harms experienced by the most marginalized human beings across the planet."[54] In other words, the social and economic systems that have created the exploitation and environmental injustices faced by humans are the same processes that have led to ecological tragedies from climate change to massive species loss. Remedies, therefore, must also include species as a "category of difference [that is] entangled in the production of environmental injustice."[55] Humans and the more-than-human world are deeply connected, both in manifestations of injustices and as agents of change. In Cascadia, where the presence of nonhuman nature has shaped legacies of extraction and displacement as well as struggles for tribal sovereignty, ecosystem protection, and climate justice, this relationship is particularly salient.

OUTLINE OF THE BOOK

This book is truly an interdisciplinary endeavor, with contributions from sociology, political science, geography, history, English, architecture, and urban planning. It is organized in three thematic parts. In "Urbanization," authors explore the historical and contemporary transformations and destruction of habitats and ecosystems brought by capitalism and urbanization; the legacy and importance of resource extraction; and the growth of regional infrastructure, built environments, and cities. In "Inequalities," authors explore Cascadia in relation to the history and contemporary status and movements for sovereignty of Indigenous peoples; the consequences of colonial and postcolonial periods; movements for environmental justice; poverty and inequality; and labor relations and struggles. In "Governance," authors explore environmental and urban democracy; the politics of states, territory, and sovereignty; the politics and dynamics of political scale (local, regional, state, national, global); green urbanism; and the impact of urban and environmental social movements. Together, the chapters build a narrative that deepens our understanding of the historical and contemporary political, economic, cultural, geographical, and aesthetic strands that make up the contradictions and possibilities of environmental justice in urban Cascadia.

NOTES

We wish to acknowledge the Coast Salish, Chinookan, and other Indigenous peoples of the territories included in this book. The authors recognize the realities of settler colonialism and the struggles of Indigenous peoples for sovereignty. These struggles are not consigned to the past. We intend this book to spark action toward a genuine politics of recognition, reconciliation, and Indigenous sovereignty throughout the region.

1 Patrick J. Smith, "Transborder Cascadia: Opportunities and Obstacles," *Journal of Borderlands Studies* 19, no. 1 (2004): 99–121.

2 Peter Berg, "Bioregionalism," Planet Drum, accessed January 20, 2020, www.planetdrum.org/bioregion_bioregionalism_defined.htm.

3 To get a sense of McCloskey's full political, scientific, and aesthetic thoughts on Cascadia, visit the Cascadia Institute website and especially look at his full color map of bioregional Cascadia at https://cascadia-institute.org.

4 "Cascadia: Land of Falling Waters," Cascadia Institute, accessed January 20, 2020, https://cascadia-institute.org/index.html#overview.

5 "Cascadia."

6 Cascadia bioregionalists even have their own flag. Known as the Doug Flag, Portlander Alexander Baretich created it in 1994. It is adorned with a Douglas fir and the colors blue for water, white for snowcapped mountains, and green for the evergreen forests. You can read more at www.cascadianow.org/about-cascadia/our-flag.

7 "About Us," *Cascadia Magazine*, accessed January 20, 2020, https://cascadia-institute.org/index.html#overview.

8 "About Us."

9 "About Cascadia and Bioregionalism," CascadiaNow!, accessed January 20, 2020, www.cascadianow.org/bioregionalism.

10 "About Cascadia."

11 Signe Marie Cold-Ravnkilde, Jaidev Singh, and Robert G. Lee, "Cascadia: The (Re)Construction of a Bi-National Space and Its Residents," *Journal of Borderlands Studies* 19, no. 1 (2004): 59–77.

12 Cold-Ravnkilde, Singh, and Lee, "Cascadia," 61.

13 Cold Ravnkilde, Singh, and Lee, "Cascadia," 60.

14 Neil Brenner, "Globalisation as Reterritorialisation: The Re-Scaling of Urban Governance in the European Union," *Urban Studies* 36, no. 3 (1999): 431–51.

15 See Matthew Sparke, "Excavating the Future in Cascadia: Geoeconomics and the Imagined Geographies of a Cross-Border Region," *BC Studies* 127 (2000): 5–44.

16 Patrick J. Smith, "Branding Cascadia: Considering Cascadia's Conflicting Conceptualizations—Who Gets to Decide?," *Canadian Political Science Review* 2, no. 2 (2008): 65.

17 Carl Abbott, *Greater Portland: Urban Life and Landscape in the Pacific Northwest* (Philadelphia: University of Pennsylvania Press, 2001).

18 Smith, "Branding Cascadia."

19 Jay Inslee, "British Columbia, Washington State to Create Cascadia Innovation Corridor to Promote Regional Economic Opportunities in Tech Industries," September 20, 2016, www.governor.wa.gov/news-media/british-columbia-washington-state-create-cascadia-innovation-corridor-promote-regional.

20 Inslee, "British Columbia, Washington State to Create Cascadia Innovation Corridor."

21 "The Future Is Cascadia," Cascadia Innovation Corridor, accessed January 20, 2020, https://connectcascadia.com/#about.

22 A good example of the latter is the Cascadia Urban Analytics Cooperative (CUAC), a joint effort by Microsoft, the University of Washington, and the University of British Columbia. The CUAC is an "applied, interdisciplinary, regional center that brings together academic researchers, students, and public stakeholder groups to address topics affecting citizens of the Cascadia region. The specific focus of CUAC is . . . high-impact data science and analytics projects that aim to understand and extract valuable actionable information out of data from urban environments in the Cascadia region at a neighborhood level." See "About," Cascadia Urban Analytics Cooperative, accessed January 20, 2020, www.cascadiadata.org.

23 Craig McCulloch, "Pandemic Takes Toll on British Columbia's Tourism, as Border Closure Extends to June 21," KNKX, May 19, 2020, www.knkx.org/post/pandemic-takes-toll-british-columbias-tourism-border-closure-extends-june-21.

24 Amanda Coletta, "Canada's Coronavirus Performance Hasn't Been Perfect. But It's Done Far Better Than the U.S.," *Washington Post*, July 15, 2020.

25 Cascadia Innovation Corridor, *Cascadia Vision 2050* (2020).

26 Erik Swyngedouw, "Apocalypse Forever? Post-Political Populism and the Spectre of Climate Change," *Theory, Culture & Society* 27, no. 2–3 (2010): 213–32; Sparke, "Excavating the Future."

27 Nik Heynen, Maria Kaika, and Erik Swyngedouw, *In the Nature of Cities: Urban Political Ecology and the Politics of Urban Metabolism* (London: Routledge, 2006).

28 Neil Brenner and Christian Schmid, "The 'Urban Age' in Question," *International Journal of Urban and Regional Research* 38, no. 3 (2014): 731–55.

29 Nik Janos, "Urbanising Territory: The Contradictions of Eco-cityism at the Industrial Margins, Duwamish River, Seattle," *Urban Studies* 57, no. 11 (2020): 2282–99; Corina McKendry, "Cities and the Challenge of

Multiscalar Climate Justice: Climate Governance and Social Equity in Chicago, Birmingham, and Vancouver," *Local Environment* 21, no. 11 (2016): 1354–71.

30 See also "White Supremacy & Resistance," special issue, *Oregon Historical Quarterly* 120, no. 4 (2019).

31 Philip Thoennes and Jack Landau, "Constitutionalizing Racism: George H. Williams's Appeal for a White Utopia," *Oregon Historical Quarterly* 120, no. 4 (2019): 468–87.

32 Shane Burley and Alexander Reid Ross, "From Nativism to White Power: Mid-Twentieth-Century White Supremacist Movements in Oregon," *Oregon Historical Quarterly* 120, no. 4 (2019): 564–87.

33 Mike Baker, Nicholas Bogel-Burroughs, and Kaitlin Gillespie, "A Day of Protest in Portland as 'Proud Boys' Converge on the City," *New York Times*, September 26, 2020.

34 Britta Lokting, "Fear and Loathing in Cascadia," *The Baffler*, February 27, 2018, https://thebaffler.com/latest/cascadia-lokting.

35 Elaine O'Connor, "B.C.'s Racist Past Spans Many Years, Many Races," *The Province* (Vancouver, BC), October 2, 2013.

36 Al Donato, "White Nationalism and Right-Wing Extremism Aren't New to Canada," CBC, accessed November 25, 2020, www.cbc.ca/cbcdocspov/features/white-nationalism-and-right-wing-extremism-arent-new-to-canada.

37 British Columbia, Office of the Human Rights Commissioner, "Hate and the Rise of White Supremacy," accessed October 7, 2020, https://bchumanrights.ca/key-issues/hate-and-white-supremacy/.

38 The *herrenvolk* in German.

39 Cold-Ravnkilde, Singh, and Lee, "Cascadia," 61.

40 Matthew Klingle, *Emerald City: An Environmental History of Seattle* (New Haven, CT: Yale University Press, 2007), 81.

41 Robert A. J. McDonald, *Making Vancouver* (Vancouver: UBC Press, 1996).

42 Richard White, *The Organic Machine: The Remaking of the Columbia River* (New York: Hill and Wang, 1995).

43 Linn Gould and B. J. Cummings, *Duwamish Valley Cumulative Health Impacts Analysis: Seattle Washington*, Duwamish River Clean-Up Coalition, March 26, 2013, http://justhealthaction.org/wp-content/uploads/2013/03/Duwamish-Valley-Cumulative-Health-Impacts-Analysis-Seattle-WA.pdf, 2.

44 "Child Poverty," Children's Alliance, accessed November 25, 2020, https://childrensalliance.org/.

45 National Council of Welfare Reports, *Poverty Profile: Special Edition*, accessed November 25, 2020, www.canada.ca/content/dam/esdc-edsc/migration/documents/eng/communities/reports/poverty_profile/snapshot.pdf.

46 Oregon Office of Forecasting, Research and Analysis, *Fall 2014 DHS OHA Regional Forecasts by District*, accessed February 2, 2020,

https://digital.osl.state.or.us/islandora/object/osl%3A90248/datastream/OBJ/view; Oregon Health Authority, *Oregon's State Health Assessment 2018*, accessed February 2, 2020, www.oregon.gov/oha/PH/ABOUT/Documents/sha/state-health-assessment-full-report.pdf.

47 Greg Halseth and Laura Ryser, "A Primer for Understanding Issues around Rural Poverty," Community Development Institute at UNBC, September 2010, www.unbc.ca/sites/default/files/sections/community-development-institute/a_primer_for_understanding_issues_around_rural_poverty.pdf.

48 See, for example, https://wallethub.com/edu/most-least-green-cities/16246; https://theculturetrip.com/europe/denmark/articles/the-10-greenest-cities-in-the-world; and www.greencitytimes.com/vancouver.

49 Harriet Bulkeley and Michele M. Betsill, "Revisiting the Urban Politics of Climate Change," *Environmental Politics* 22, no. 1 (2013): 136–54.

50 Vanesa Castán Broto, "Urban Governance and the Politics of Climate Change," *World Development* 93 (2017): 1–15.

51 Melissa Checker, "Wiped Out by the 'Greenwave': Environmental Gentrification and the Paradoxical Politics of Urban Sustainability," *City & Society* 23, no. 2 (2011): 210–29; Jennifer L. Rice et al., "Contradictions of the Climate-Friendly City: New Perspectives on Eco-gentrification and Housing Justice," *International Journal of Urban and Regional Research* 44, no. 1 (2020): 145–65; and Miriam Zuk et al., "Gentrification, Displacement, and the Role of Public Investment," *Journal of Planning Literature* 33, no. 1 (2018): 31–44.

52 Winifred Curran and Trina Hamilton, eds., *Just Green Enough: Urban Development and Environmental Gentrification* (London: Routledge, 2018).

53 David Schlosberg, *Defining Environmental Justice: Theories, Movements, and Nature* (Oxford: Oxford University Press, 2007).

54 David N. Pellow, *What Is Critical Environmental Justice?* (Cambridge: Polity Press, 2018), 2.

55 Pellow, *What Is Critical Environmental Justice?*, 17.

PART 1

URBANIZATION

CHAPTER 1

DWELLING WITH THE ENTWINED ECOTOPIAN AND TECHNO-UTOPIAN LEGACIES OF CASCADIA

JEFFREY C. SANDERS

DURING THE MID-1960S, A LANDSCAPE ARCHITECT NAMED Richard Haag envisioned a park on the site of a dilapidated coal gasification plant near Seattle's Lake Union. He made a controversial choice. Rather than erase evidence of the site's toxic past, he left the turn-of-the-last-century smoke stacks, gas tanks, refinery equipment, metal ladders, and catwalks in place. When the park debuted at the height of the postwar environmental movement, it offered the ultimate adventure playground in that era just before Americans became so litigious about head injuries and heavy metals. In a masterstroke, Haag, the founder of the University of Washington's landscape architecture school, left a stark cluster of sinister towers, a sculptural reminder of the city's industrial-era sins surrounded by a sea of green grass (see figures 1.1 and 1.2). As a budding seven-year-old environmental historian at the time, I absorbed his commentary on the relationship between nature and technology and what I can now see is Seattle's tortured relationships to each.[1]

Gas Works Park epitomized Seattle's ecotopian moment. It opened in 1974, a year before Ernest Callenbach published his influential *Ecotopia*, the story of a journalist's journey to report on a bioregional secessionist movement in the Pacific Northwest. Berkeley-based Callenbach's tale might have struck some Seattle readers as yet another

FIGURE 1.1. Gas Works Park, Seattle, Washington. Richard Haag made the industrial ruin a centerpiece of the green space. Photo by WikiPendant at Wikimedia Commons. Licensed under CC BY-SA 4.0. https://en.wikipedia.org/wiki/Gas_Works_Park#/media/File:Seattle_Gas_Works_Park_old_gas_plant2013.jpg.

FIGURE 1.2. Aerial view of Gas Works Park on the shores of Lake Union. Photo by Liesl Matthies. Licensed under CC BY-SA 3.0. https://en.wikipedia.org/wiki/Gas_Works_Park#/media/File:Aerial_Gas_Works_Park_September_2011.jpg.

imperial Californian intrusion. But utopias are always fraught impositions. *Ecotopia* definitely captured the mood as burned-out baby boomers from the 1960s turned their energies to salvaging ruins in the age of ecology. His fiction wasn't that far off either. A more chastened, restoration-minded, post-industrial urbanism seemed close at hand after Boeing's boom busted and the oil crisis crept in. Visible from the I-5 express lanes, Gas Works Park's rusting skeletons offered a glimpse into a post-oil future and a reminder that residues of the past are never past, that today's technological enthusiasm may end in ruin.[2]

Like *Ecotopia*, the steampunk green space tapped utopian aspirations predating postwar environmentalism or deindustrialization. More than most North American metropolises, Seattle and the hinterland it claims have been long consumed by techno-utopian and arcadian visions of ecotopia tethered together in an uneasy embrace. For more than 170 years, the region endured a succession of these fantasies, rapidly cycling moods paralleling economic busts and booms. This chapter emphasizes two main utopian drives that rest on the shaky foundations of the Cascadia subduction zone. To a large degree, the history of the Emerald City entrepôt and the region that pays it tribute are held together by visions of techno-utopia (the gasworks) and ecotopian aspirations (Haag's uneasy harmony of machine and green). These impulses are entwined, though they often seem in open conflict, like the park's visual contradictions.[3]

EARLY UTOPIA AND ECOTOPIA

As Haag and Callenbach suggest, utopian inspiration may come in literary or physical forms. Utopia—literally "no place"—has been attached to every idealized place or marketing scheme since Thomas More first published his original utopian work in 1516. *Utopia* dealt with specific concerns of More's moment when Renaissance humanists grappled with recovered Greek texts, their pagan power upending the status quo. As historian Stephen Greenblatt notes, More's *Utopia* created a "visionary, detailed blueprint," or at least a literary one, a way to imagine a context for ideas that he and his contemporaries encountered.[4] More's *Utopia* described abolition of private property, benefits of communal ownership, and the evils of concentrated wealth, making his book a perennial favorite among subsequent utopians. His idealized island commonwealth became a sensation, soon translated from Latin into French, Dutch, and English.[5]

From the start, visions of utopia tangled with material places and people that Europeans, armed with new navigation tools and faster ships, encountered in the New World. In fact, *Utopia*'s themes drew directly from such encounters. Amerigo Vespucci's voyages in the early 1500s and his subsequent writings made their way into More's imagination and narrative. During the same period, explorer Giovanni da Verrazzano dubbed most of the northeastern Americas Acadia, a variation on ancient Arcadia, the mythical pastoral region where the god Pan roamed. Subsequent urban planning owes elements to these utopian projects too. In fact, More's vivid depiction of gardens, clean water, and "commodious and handsome" streets in Utopia's capital city—unlike his fetid London—inspired many New World settler colonial projects. *Utopia* in turn influenced hundreds of expansionist schemes to follow in the "age of discovery," a feedback loop that continues to run.[6]

Green regionalism such as Callenbach's is no less rooted in this colonial history and the inequalities it produced. Historian Richard Grove argues in *Green Imperialism* that utopia and western European environmentalism emerged in tandem too. South Sea Islands that French imperialists visited and claimed shaped European environmental imaginations. Islands became the "speculative and Utopian location for the atavistic 'discovery' of idyllic societies," Grove explains.[7] These bounded spaces served as a stark measure of European environmental degradation and plunder. As with More's island Utopia, real and imagined outposts provided a stage where utopian and dystopian dramas played out together.

The point is that our enduring desire to perfect models of ideal social and environmental relations took hold early, tangled with the feverish context of empire. *Utopia* crystalized and widely circulated these building blocks, linked tropes of utopias and arcadias, hazy places off the map's edge.

Contemporary Cascadian versions of green utopianism can be traced in part to nineteenth-century versions of these impulses too. Royal Horticultural Society botanist David Douglas first used the name Cascades during the 1820s when he investigated the Northwest.[8] Like his famous contemporary Joseph Banks, Douglas collected, explored, and reassigned names to places and plants. In his journal, he describes working with Indigenous people to gather their intellectual property and the biological material of their homelands: "I reached my long-wished *Pinus* (called by the Umpqua tribe *Natele*), and lost no time in examining and endeavoring to collect specimens and seeds."[9] The Douglas fir–emblazoned flag of contemporary Cascadia is named for the botanist who played this role in a larger project of place-making and taking.[10]

Regional visions are not merely an exercise of European imagination. Indigenous regional ties of kinship, trade, and shared ecologies long predated colonial impositions or political boundaries since time immemorial. Northwest regionalism is not only bound together by mossy forests and gray-blue Salish Sea waterways either. The Columbia River, flowing through arid inland plateaus, mountain ranges, and wet forests to the Pacific made gathering, movement, and exchange a cosmopolitan, cross-mountain, regional affair.[11] The Cascadia "bioregion" rests on a succession of these projections—from the Oregon Country that US, UK, and Indigenous people fought to control, to Jefferson's soulless grid, to destructive reservation lines that Territorial Governor, Superintendent of Indian Affairs, and Pacific Railroad surveyor Isaac Stevens measured and staked. This corner of the continent has been contested among empires and sovereign Nations ever since. These utopian and regional imaginings leave behind material evidence—lines, ruins, and landmarks worth examining.

STEAMPUNK UTOPIA

If regional and utopian visions have been with us for a while, their uniquely techno-utopian and ecotopian iterations manifested at a frightening pace between 1870 and 1910, imposing an enduring set of meanings across the region, from periphery to the metropole. The period especially frames much of our thinking about city and hinterland.

Not long after Douglas tromped through Northwest forests, Americans arrived to cut them down and make property in their place. Early Seattle colonists called the land where they landed "New York by and by," a suggestion of no place from the start. The romance of skid roads and gold rushes dominates popular narratives of

Northwest urbanism. Mud-filled streets, sewage disasters, and the cleansing fire of 1889 round out Seattle's popular origin story, a dystopian picture of hapless western disorder dished out to tourists on the Seattle Underground Tour.[12] However alluring the history of gold rushes and lumberjacks may be, coal also was a crucial factor in kicking modern dreams of regional imperium into high gear. Wood was still the dominant cash crop, but the arrival of steam engines and private capital quickened exploitation of the largest coalfields on the West Coast.[13]

Seattle's hinterland was no Appalachia, yet the first local railroad in the Seattle area, built in 1873 by Seattle Coal & Transportation Company, hauled coal not people or wood from mining towns around Lake Washington. Fossil fuels that miners dug in hinterland towns such as Black Diamond, Coal Creek, and New Castle made city and region cohere quickly, an industrial landscape bound more by steel tracks, railroad grades, and timetables than any cartography of hope. The gasworks arrived on Lake Union's shores in 1906, at the latter end of this era, ready to transform coal into gas to light Seattle's bungalows. Coal and steam engines made plying waters and crossing the Cascades easier. Steam donkeys dragged logs out of forests to locomotives that hauled them down to waterfront docks and then onto steamers bound for California. Each year the Great Northern Railway advertised the region and brought yet more people to it. Between 1870 and 1910, Seattle's population grew from 1,151 people to over 230,000.[14]

City engineer Reginald Heber Thomson did most to realize the techno-utopian dream of networking Seattle and its hinterland. In search of Seattle's "natural advantage," nineteenth-century engineers such as Thomson harnessed steam power to affirm visions of harmony and property rights. Unlike More's utopians, Thomson wasn't interested in abolishing private property. He created more of it. In his memoir, the midwestern transplant wrote, "Seattle was in a pit," and to "get anywhere we would be compelled to climb out if we could."[15] His oft-quoted sentiment had as much to do with making the city as to do with binding Seattle's fortunes to its hinterland.[16] Thomson hoped to efficiently move goods and maximize flows of resources, to join hinterland and city in a great symphony of extraction, water, and energy. Thomson began his career working with F. H. Whitworth, who, he recalled, "wished to make Seattle the coal field of the

Northwest." Laying low Seattle's hills allowed for easier movement of coal and other goods from the east to west side of the lakes and along gradual grades to the waterfront docks.[17]

With machines these techno-utopians planned their ideal commonwealth and its corresponding environment on the ground and in real time. If the ground wouldn't cooperate, they'd tear at it and reshape it until it did.[18] Steam shovels, locomotives, horses, and hands displaced a total volume of approximately 75,036,595 cubic yards of dirt, gravel, and rock between 1903 and 1928.[19] Thomson and his laborers flattened hills, filled tideflats, dug a ship canal between two inland lakes (Lake Washington's levels dropped nine feet in the process), linked fresh and salt water with locks, and made way for the Great Northern Railway. The engineer is infamous for sweeping away a massive mound of glacial-deposited clay and dirt called Denny Hill near downtown. Photographs from the time depict chaotic scenes of spite houses on dwindling hilltops and wooden walkways teetering over gullies. Thomson's visions of progress and improvement ultimately displaced poor people, a constant refrain in Seattle's history. He undermined the already precarious position of Indigenous Seattleites by washing gravel and mud into tideflats that supplied subsistence, while the ditches he dug unearthed ancient shell middens and the stone hearths of their ancestors.[20]

Utopia is often borne of dystopian means. Nineteenth-century techno-utopians and ecotopians were often one and the same. Violence that displaced dirt, people, and ancestral sites also cultivated the bed of ecotopian yearnings. In other words, Haag's landscape isn't as discordant as it may seem. Technology furnished the frame for our visions. Engineers and planners such as Thomson made it easier to move through hills, between waters, and along tracks that knitted together picturesque nature and urbanism, compressing experiences of time and space.[21] A landscape sleight of hand, these efforts allowed compartmentalizing too. Lake Union became a "working lake," Lake Washington destined for leisure, distant cutover forests a memory.

The nationally famous Olmsted Brothers landscape architecture firm was integral to these illusions of displacement. They created systems of unifying parkways and public spaces that did most to define districts and blend technological know-how and environmental design. Combined with Thomson's work, the Olmsteds defined Seattle's modern image and relationship to its region, training the newcomer's gaze toward vistas and away from costs.[22] The automobile and clamor for good roads only reinforced an image of Northwest cities balanced between urbane and raw. New roads linked cities ever closer to pleasure grounds such as Mount Rainier National Park in one direction and Olympic National Park in another.[23]

Seattle's first techno-utopian and ecotopian age culminated at the Olmsted-designed Alaska-Yukon-Pacific Exposition (AYPE) fairgrounds in the summer of 1909 (though the regrades were still far from finished). Seattle boosters championed the regional coming-out party, carving a late-Victorian fantasy into receding forests near Portage Bay. Historian Robert Rydell argues that the fairs of this era encapsulated the utopian and imperial drives of America's nervous ruling class: "At a time when the American economy was becoming increasingly consolidated and wealth generated by the country's economic expansion concentrated in fewer hands, the exposition builders promised that continued growth would result in eventual utopia."[24]

Souvenir bird's-eye picture postcards capture the ordered organization of streets, environmentally and racially segregated spaces, social Darwinist hierarchies, a scaled version of what white elites hoped the city, region, and world could be. In that heady moment of empire as the US consolidated its hold over the Philippines after the Spanish-American War, the AYPE touted the city's role in tying together a regional hinterland that now, boosters said, stretched from Alaska to the Pacific World.[25]

It would take at least another decade, but Denny Hill would finally be but a memory, washed to the bay. But both the gasworks and the fairgrounds remain, just down the shore from each other. Gas Works Park reminds us of Seattle's first techno-utopian reach into the hinterland, while the powerful "Rainer Vista" at the fairground continues to frame our regional view. The subsequent design for the University of Washington campus built on this original blueprint (see figure 1.3).

LABOR AND LAND

As colonial rationalizations merged with ecotopian and techno-utopian blueprints at the AYPE, not everyone was having it. When the fair opened in 1909, the Washington State Federation of Labor

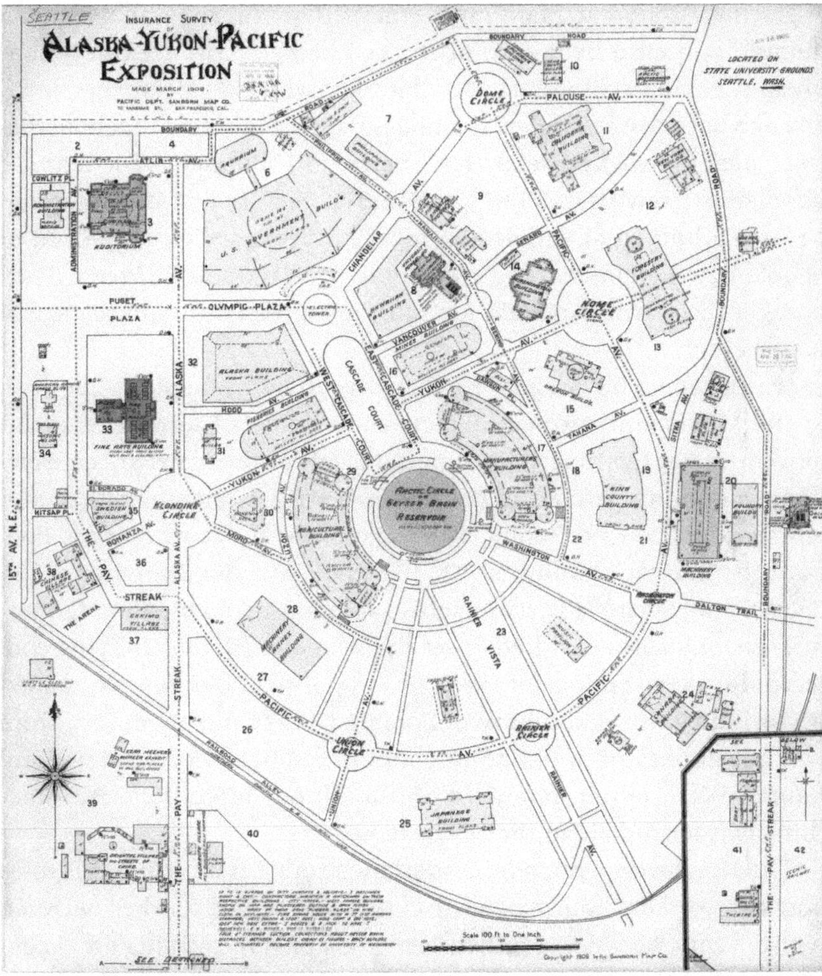

FIGURE 1.3. Alaska-Yukon-Pacific Exposition (AYPE), 1909. This image depicts the beaux arts layout of the fairgrounds (later to become the basis for the plan for the University of Washington campus). The layout reiterated the segregation of center from periphery, placing "human zoos" of people brought from new US imperial outposts of occupation. Sanborn Fire Insurance Map from Seattle, King County, Washington. Sanborn Map Company, March 1909. Map, Library of Congress, Geography and Map Division. https://www.loc.gov/item/sanborn09315_032/.

called for a boycott in opposition to the fair's open shop policy. In the wake of more than two decades of shambolic regrades and local battles between labor and capital, the region's working class was fed up.[26]

Northwest utopian plans often emerged in tandem with dystopian moments, bound by a similar logic. The rails and steamers that brought capitalists to invest also brought workers, organizers, seekers, and back-to-the-landers to build an opposing vision that nonetheless relied on tracks already laid. Between 1900 and 1920, Seattle's population grew from 80,671 to 315,312. Seattle and its hinterland became a hotbed of activism, a highly diverse social utopia full of people hewing closer to egalitarian lessons in More's *Utopia*. They brought their own templates for a just and green future in the age of the machine.

Historian James Gregory's exhaustive digital Mapping American Social Movements project depicts the variety of American social movements at the time.[27] The upper far left of the maps are conspicuous, blooming patches of color where strikes, memberships, and newspapers had a per capita concentration rivaling any other region, even the industrial Midwest. By 1912, Washington State had 6,326 dues-paying Socialist Party members—a thousand more than California and only a few hundred fewer than Ohio or Illinois. Defying contemporary conceptions of rural/urban political divides, Northwest hinterlands also had plenty of Socialist Party members too, from commissioners in Spokane to councilmembers in Aberdeen and Camas, Washington. In an era of tumultuous labor politics, something was fecund about the Northwest.

The Northwest's regional hinterlands roiled with brutal disputes between workers and the owner class's vigilantes. Clashes between workers can't be ignored either. They took on a racialized and violent form early with the Chinese Exclusion Act in 1882 and subsequent Knights of Labor–led terrorism against Chinese laborers in communities on the periphery and at the center of early Cascadia. Later in this era, mobs of white laborers drove Asian workers onto the sea and the rails, much as they had the original Seattleites. In Bellingham just two years before the AYPE, similar racial violence had targeted South Asian workers in lumber mills there.[28] It was no utopia for people of color, even if the region was alive with the spirit of "one big union."

Many radicals founded communal experiments not far from mining districts and cutover forests. According to historian Charles Pierce LeWarne, these northwestern utopians merged "a longstanding communitarian tradition with recent economic and political reform movements in the United States."[29] Some were inspired by early-nineteenth-century experiments such as Robert Owen's New Harmony. Others were Henry George–inspired single taxers who feared concentration of wealth and land speculation. Still others hoped to live out visions of socialism that they found in Edward Bellamy's *Looking Backward, 2000–1887,* the bestselling utopian novel of the era. These utopians each left their mark up and down the Salish Sea and beyond in communities they called "colonies" with names such as Freeland, Harmony, Equality, and Home—inspiration for their 1970s ecotopian descendants.[30] Yet as federal authorities tightened the screws on dissent during World War I and in wake of the Bolshevik Revolution, the First Red Scare squelched many of the era's strides. The period crested in 1919 with the Seattle General Strike.[31]

FROM HOOVERVILLE TO SPACE NEEDLE

The Great Depression and World War II utterly transformed regional relationships of work, capital, and nature again. In Seattle, the "dirty 30s" created a dystopia of photogenic Hoovervilles only recently rivaled by the post–Great Recession housing crisis in West Coast cities. With their own combination of utopian and colonial visions, regional planners during the New Deal and after responded with a new set of techno-utopian and ecotopian plans that emphasized hinterland developments.[32]

The Northwest Regional Council, a multistate planning agency, invited famed critic Lewis Mumford to consult on their planning efforts in 1939. His resulting memorandum offered an eloquent critique aimed at boosters in previous decades and "the Wastes, the extravagance, the misuse of our natural resources." He countered with "regional rehabilitation" that he thought should focus on "the best development of the Columbia River Valley."[33] Mumford built his bioregionalism on a nineteenth-century scaffolding (just as subsequent ecotopians and Cascadians would), attempting to heal the regional cleavage he saw at the banks of the Columbia, dividing Oregon and Washington.

A prescient critic of the carbon economy, Mumford championed hydroelectric regionalism, believing it a more democratic and decentralizing option than coal. Subsequent federal dollars bound the region with transmission towers and power lines. A "regional renewal," Mumford noted, could remake cities such as Seattle as well, offering "the reforestation of urban culture [and] the prevention of social erosion."[34] This regional cohesion, however, came at the expense of habitat and access to fish at "usual and accustomed places," such as Kettle Falls, that dam builders submerged.[35]

Meanwhile, on Seattle's Profanity Hill overlooking Pioneer Square, the Seattle Housing Authority used New Deal dollars to build the country's first integrated public housing developments in a "blighted" section of the city's Japantown neighborhood. It was a utopian and progressive gesture but one that also resulted in cultural "erasure," as historian Megan Asaka notes. It was a blow for a community on the eve of a war that would further devastate Seattle's Japanese communities, or Nihonmachi. Federal Housing Administration policies during the depression and after would generally favor whites in an enduring pattern that shaped real estate markets well into the 1960s.[36]

Rather than knit together city and decentralized communities, New Deal grants and hydroelectricity (along with petroleum and eventually superhighways) pulled core and periphery apart. After the war, an emergent military-industrial complex rested on cheap energy and federal investment to promise endless growth and jobs. "A job at Boeing's," baseboard-heated homes, and new suburban shopping malls cemented this "consumer's republic" in place. At midcentury, grandchildren of free lovers, stump farmers, and coal miners followed low-interest mortgages to racially segregated dystopias with un-ironic names such as Lake Forest Park or Broadview. They chewed into forests and stolen Japanese American truck farms, exacerbated stormwater runoff, and leaked detergents and human waste into nearby streams and lakes.[37]

In an effort to buoy their sagging fortunes amid the exodus, downtown boosters turned again to utopian displays to match the suburbs. The high-profile Century 21 Exposition in 1962, also known as the Seattle World's Fair, was their answer to sprawl. Disney Inc. consulted on the design. Haag even cut his landscape design teeth at the fair. Unlike the 1909 expo situated on Seattle's early crabgrass frontier, downtown boosters now worried that the center might not hold. This time the productive marriage of federal largesse and heavy-hitting national brands (Boeing, General Electric, and IBM) bankrolled a utopian vision that underscored Seattle's place in a consolidating national culture (see figure 1.4).[38]

FIGURE 1.4. Century 21 Exposition publicity. The 1962 fair expressed the promise of a space-age utopia, inspiring a generation of baby boomers, such as Microsoft's cofounder Paul Allen. MOHAI, Walter Straley Collection of Century 21 Photographs, 1965.3598.26.91.

As Cold War counterinsurgencies heated up in Southeast Asia and the space race accelerated, the missile gap and regional fortunes combined, propelled by hydropower, aluminum, and jet fuel. Empire was intrinsic to this modern fair, too, and the client-state economic relationships it celebrated. Seattle's skyline and collective imagination has yet to outlive the moment. The era's beloved relic is the midcentury floor lamp Space Needle, still reaching toward Kennedy's New Frontier.[39]

Amid Cold War exuberance, some regional elites took a moment to reflect on problems unbridled growth actually produced. These technocrats hoped to balance growth and quality of life. In parallel with the fair, civic leaders such as Jim Ellis, a municipal bond lawyer, created more sober visions. Their creation—the Municipality of Metropolitan Seattle (METRO)—addressed regional growth's dark side coming into focus at midcentury. METRO succeeded at the polls and is still celebrated as a pioneering example of enlightened regional planning, even as its class-based plans ultimately displaced pollution and made way for more exurban development.[40]

Meanwhile, with urban renewal dollars in hand, downtown developers persisted in shedding the blight of history altogether. Like tourists smashing crab shells at the waterfront seafood shacks, their state-sponsored wrecking balls laid waste to the old single-occupancy hotels that housed the working poor, making way for generous multilevel parking garages in the city's core. Century 21 presented another model for the continued headlong rush into "no place."

ECOTOPIA AND DEINDUSTRIALIZATION

As ecotopian dreams rose again, they conflicted with the George Jetson futurism of Century 21. At the end of the 1960s, the word *ecology* offered a soothing metaphor. From the Greek *oikos*, it suggested making a home in the world, a scale easily understood by youth born of a million basement rumpus rooms. For activists it also captured the era's interrelated environmental and social problems, from Agent Orange to "the urban crisis" to the looming energy shortage. Ecology evoked its own utopian images of young people tending terrariums, playing with giant earth balls in the park, or maybe folk dancing—certainly not space-age, precision-machined metal tubes.

Yet the energy crisis and Seattle's "Boeing Bust" did the most to underline the region's fragile economic model. Slogans such as "Small Is Beautiful" and "Do It Yourself" characterized the emerging ecotopian ethos: the perennial wish for a more human and regional scale.

As Boeing cut jobs by half beginning in 1971 and a workforce sold its homes, would-be ecotopians leapt into the breach left by deindustrialization. A generation of back-to-the-landers, *Whole Earth Catalog* readers, "tool freaks," organic yogurt makers, and savvy hip capitalists trickled into the region to replace the Boeing workers.[41]

Callenbach, the interloper from Berkeley who wrote *Ecotopia*, merely captured a mood in the air along the new I-5 corridor. In his second-rate futurist, utopian book, Callenbach drew, like so many before, straight from More's *Utopia*. His breakaway region roughly matched the outlines that today's Cascadian enthusiasts and yesterday's Oregon Country traced too. He also imagined the region as racially homogenous. As "families of great wealth fled" the region, he fantasized, ecotopians took over the factories and farms. Citizens daylighted streams for salmon and banned automobiles. His vision resembled the highest hopes of contemporary urban and rural dwellers in Portland, Seattle, and Vancouver in the 1970s. In town, ecotopians began a process of gentrification, taking over low-rent bungalows and creating group houses in first-ring neighborhoods such as Fremont, Central District, and Capitol Hill, joining a neighborhood-based "middle city" political renaissance underwritten partly by War on Poverty dollars.[42]

The real ecotopians never called themselves by the name, of course. But groups devoted to "steady state ecology" (the notion of maintaining an equilibrium in an ecosystem), such as Northwest Tilth, advocated alternative agriculture from the Willamette Valley to the Okanagan in British Columbia. In a decade of rapid consolidation in agriculture and rising rural bankruptcies, ecotopian groups reclaimed small farms in Seattle's nearby hinterland and learned to grow winter vegetables and convince others to eat them.[43]

Borrowing from the example of Puget Sound utopians in the past, the ecotopians of the 1970s did more to transform the way people in the region ate and thought about food security, energy, and waste than any utopian writer could imagine. In hinterland places such as the "Magic Skagit," where Gene Kahn of Cascadian Farms made organic blueberries mainstream, or by infiltrating city departments to create composting and recycling programs, or by creating a municipal community agriculture program, they made ecotopian visions a reality bit by bit.[44] If there was an emblematic ecotopian counter space to match the earlier fairs, it was the sagging Pike Place Market.

In the late 1960s, neighborhood activists and preservationists battled downtown commercial boosters who wanted to demolish it.[45]

On the surface it appeared myriad DIY efforts of preservationists and counterculture idealists had salvaged the city and region on their own, just like Callenbach's secessionists. But many of these urban innovations, despite the tight-belted 1970s, rested on a package of publicly financed planning initiatives called Forward Thrust (an elaboration of METRO). By the 1980s, enjoying expanding green space such as Gas Works Park, cleaner water, and better services, Seattleites promptly forgot that it all hinged on public investment.[46]

TECHNO-UTOPIAN DISCONTENTS

Bill Gates and Paul Allen launched Microsoft the same year Callenbach published his regional fantasy.[47] Unlike previous utopian pushes that profoundly remade physical landscape, the PC and software revolution seemed benign, clean. Microsoft even located its "campus" on Seattle's greenfield edges in the 1980s. As the city crawled from the doldrums and into its next flannel phase, it was no place for the industry's antiseptic image. The tech economy, of course, is no cleaner than the benzene and PCBs that leaked from Gas Works into Lake Union. The social costs also accumulated. In our latest disembodied techno-utopian reveries, we might convince ourselves that we are liberated from place altogether were it not for the social ruin we see when the VR goggles come off. The conspicuous connection between the digital economy and contemporary economic inequality and environmental despoliation in Cascadia is now hard to deny.

Yet Microsoft's rising fortunes soon attracted cofounder Allen back to the old Century 21 fairgrounds, like a spawning chinook. (The Bill and Melinda Gates Foundation followed later.) Allen led the gentrifying charge that sociologist Richard Florida later dubbed the "creative class," drawn to ecotopian-nurtured amenities and urban authenticity. By now we are familiar with the colonizing influence and imperial reach of the leading tech companies (especially the personal data they mine from us). But these companies that seem to hover in the clouds have had serious and specific effects on the urban and natural environments they and their employees actually inhabit—from energy-consuming data farms on the Columbia River to e-waste to tent cities near the highway.[48]

By the mid-1990s, Allen purchased swaths of cheap land on South Lake Union anticipating future developments. He built a museum to house his popular culture collection in the Needle's shadow. Jeff Bezos and Amazon soon followed, choosing an urban rather than "campus" model for the rapidly growing company. For years Amazon inhabited the repurposed Marine hospital on Beacon Hill in true ecotopian form. But as the company grew, it too moved closer to Allen's neighborhood of shiny office buildings and rooftop garden condos.

In the first decades of the actual twenty-first century, the high-tech industry has upended real estate markets, driving affordable housing to the margins. During 2015 to 2017, Seattle's median home values increased from $538,000 to $708,000.[49] Many critics have taken aim at high-tech companies such as Seattle-based Amazon. The richest man in the world, Bezos, symbolizes for many the wrenching transformations, becoming a target for grassroots groups, city councilmembers, and telephone pole poster mockery (see chapter 9).

Every era has its landmarks and touchstones that suggest utopian ambitions, such as the fairgrounds and the Space Needle. The "Amazon Spheres," erected as part of Amazon's new headquarters in Seattle's Denny Regrade neighborhood, may be ours for the digital age, a confluence of digital and green oligarchic escape fantasies (see figure 1.5). The four-story tall terrariums—containing 40,000 plants from "cloud forest regions of over 30 countries"—look like a hippie's Buckminster Fuller fever dream mashed up with a Kew Gardens greenhouse. The building suggests a link between the latest tech bros and techno-utopians and their shared counterculture and imperial roots. The promotional website notes, "The Spheres are a result of innovative thinking about the character of a workplace and an extended conversation about what is typically missing from urban offices—a direct link to nature."[50] The Spheres sit in the middle of a Corbusier-like super bloc style development built by the company that embodied the city's latest confluence of green and techno-utopian dreams. The symbolic orbs make a monument to greater wealth inequality in the midst of this latest utopianism. And just as Seattle became a center of resistance in the nineteenth-century, the city again leads the national conversation.

Resistance to this tech imperial vision grabbed headlines in April 2018 when city councilor Kshama Sawant addressed a crowd in front of the Spheres. More than a hundred years after Denny Hill

FIGURE 1.5. The Amazon Spheres. The Spheres evoke the enduring pairing of greenhouses and colonialism, from Kew Gardens herbariums to interplanetary utopianism. Photo by Joe Mabel. Licensed under CC BY-SA 4.0. https://commons.wikimedia.org/wiki/File:Amazon_Spheres_05.jpg.

came down, Sawant spoke truth to power in a different moment but on the same site beneath a phantom hill. That year, Sawant led the drive to tax tech employers $235 for each new employee hired in an effort to address the city's housing crisis. The cantankerous Socialist was there to throw rhetorical rocks at the glass house behind her with cameras rolling. Defending the council's proposal, she noted it would be "pocket change for corporations like Amazon . . . Pocket change for these billionaires."[51]

Less than a year later, in Seattle's most expensive city election ever, Sawant was able to hold on to her seat on the city council after a grueling campaign season in which the Amazon Corporation and the Seattle Chamber of Commerce spent at least $5 million to unseat her and other supporters of the head tax. In the same election season, as voters in Seattle affirmed the work of Sawant and her colleagues, voters in the exurban hinterland supported ballot measures reducing taxes that would have helped to shore up transportation infrastructure in the region. In an era of economic inequality and concentrated wealth that now surpasses the Gilded Age, the electorate is once again grappling with dystopian realities, while exurban voters chide Seattle for being a "socialist utopia."

If the Century 21 Exposition's celebration of warfare capitalism fired youthful imaginations of our contemporary techno-utopians, it is useful to recall that in the 1950s and 1960s, when boomers such as Gates, Allen, and Bezos were growing up, the top marginal federal tax rate hovered between 92 percent and 91 percent. In 2018, it was around 37 percent.

Politicians such as Sawant, whose demands some voters find so strident today, echo the booming voices of street corner Wobblies. But she actually has more in common with middle-of-the-road municipal planners of the 1950s and 1960s than some would like to admit. If you don't believe me, take a look at the laundry list of public spending—from subways to the Kingdome and zoos to sewer treatment facilities—in the METRO and Forward Thrust pamphlets of that era, when technocratic regional elites proposed civic improvements worth $356 million. Their demands read like a Green New Deal for postwar Seattle. It was. We owe Gas Works Park to these public funds.[52]

Gas Works Park and the Amazon Spheres pair well, I think. Recently, to the horror of some old Seattle boosters, a local journalist invited a public vote on whether the Spheres might replace the Space Needle as Seattle's updated futurist emblem. But Gas Works makes us pause and consider the limits of our technological enthusiasm and the carbon economy with which, despite ecotopian efforts, we must still reckon. Rather than no place or every place that the Spheres embody, the sooty towers speak to the reality of the place beneath our feet and our responsibility to dwell on our environmental history and ever present. The lakeside ruins suggest that public space and public memory can be central to urban life. The Amazon Spheres, in contrast, private and contained, form a closed system. They inspire interest in ecotopian and techno-utopian dreams, but they could be anywhere, maybe on Mars or wherever Amazon plans to go next.[53]

But if we are looking for symbolic spaces and icons that speak to Seattle's special blend of the ecotopias and techno-utopias or to the problems we face in the Anthropocene or the Carbonocene or the Capitalocene or whatever we call it now, my vote is for Gas Works.

We could do worse. Haag even built a massive hill there from the site's polluted soils, perhaps a comment on the loss of Denny Hill? It's a great place to put down your devices and go fly a kite.

NOTES

1 William S. Saunders, ed., *Richard Haag: Bloedel Reserve and Gas Works Park* (New York: Princeton Architectural Press, 1998), 6–28.

2 Ernest Callenbach, *Ecotopia: The Notebooks and Reports of William Weston* (New York: Bantam Books, 1990).

3 For histories of Seattle, see Coll Thrush, *Native Seattle: Histories from the Crossing-Over Place* (Seattle: University of Washington Press, 2007); Matthew W. Klingle, *Emerald City: An Environmental History of Seattle* (New Haven, CT: Yale University Press, 2007); and Jeffrey C. Sanders, *Seattle and the Roots of Urban Sustainability: Inventing Ecotopia* (Pittsburgh: University of Pittsburgh Press, 2010). For western utopian urbanism, see John M. Findlay, *Magic Lands: Western Cityscapes and American Culture after 1940* (Berkeley: University of California Press, 1992); Kate Brown, *Plutopia: Nuclear Families, Atomic Cities, and the Great Soviet and American Plutonium Disasters* (New York: Oxford University Press, 2013); and James C. Scott, *Seeing Like a State: How Certain Schemes to Improve the Human Condition Have Failed* (New Haven, CT: Yale University Press, 1998).

4 Thomas More, *Utopia* (Boston: Bedford/St. Martin's, 1999); and Stephen Greenblatt, *The Swerve: How the World Became Modern* (New York: W. W. Norton, 2011), 230.

5 The Greek word *utopia* means "no place." More, *Utopia*, 15.

6 Vasco de Quiroga, Bishop of Michoacán in the 1530s, built several communities in colonial-era Mexico based on More's ideas. More, *Utopia*, 6–8; and Greenblatt, *Swerve*, 229.

7 Richard H. Grove, *Green Imperialism: Colonial Expansion, Tropical Island Edens and the Origins of Environmentalism, 1600–1860* (Cambridge: Cambridge University Press, 1995), 223.

8 Jack Nesbit, *The Collector: David Douglas and the Natural History of the Northwest* (Seattle: Sasquatch Books, 2010).

9 David Douglas, *Journal Kept by David Douglas during His Travels in North America 1823–1827* (London: W. Wesley and Son, 1914), 230, www.sos.wa.gov/library/publications_detail.aspx?p=56.

10 Nesbit, *Collector.*

11 For a deeper history of region in the Pacific Northwest, see Lissa K. Wadewitz, *Nature of Borders: Salmon, Boundaries, and Bandits on the Salish Sea* (Seattle: University of Washington Press, 2012); Richard White, *The Organic Machine: The Remaking of the Columbia River* (New York: Hill and Wang, 1995); Andrew H. Fisher, *Shadow Tribe: The*

Making of Columbia River Indian Identity (Seattle: Center for the Study of the Pacific Northwest and University of Washington Press, 2010); and Thrush, *Native Seattle*, 105–25.

12 Roger Sale, *Seattle, Past to Present* (Seattle: University of Washington Press, 1976); Murray Morgan, *Skid Road: An Informal Portrait of Seattle*, rev. ed. (New York: Ballantine Books, 1971); and William C. Speidel, *Sons of the Profits, or There's No Business Like Grow Business: The Seattle Story, 1851–1901* (Seattle: Nettle Creek Publishing Company, 1967).

13 Lisa Mighetto, Marcia Babcock Montgomery, and United States National Park Service, Columbia Cascades Support Office, *Hard Drive to the Klondike: Promoting Seattle during the Gold Rush: A Historic Resource Study for the Seattle Unit of the Klondike Gold Rush National Historical Park* (Seattle: National Park Service, Columbia Cascades Support Office, 1999).

14 Jennifer Ott, "Seattle Coal & Transportation Company Begins Operating Seattle's First Railroad on March 22, 1872," HistoryLink.org, October 10, 2012, www.historylink.org/File/5412. For coal in the West, see Thomas G. Andrews, *Killing for Coal: America's Deadliest Labor War* (Cambridge, MA: Harvard University Press, 2008); Richard A. Rajala, *Clearcutting the Pacific Rain Forest: Production, Science, and Regulation* (Vancouver: UBC Press, 1998); and William Cronon, *Nature's Metropolis: Chicago and the Great West* (New York: W. W. Norton, 1991), 74–93.

15 Reginald Heber Thomson, *That Man Thomson* (Seattle: University of Washington Press, 1950), 14.

16 William H. Wilson, *Shaper of Seattle: Reginald Heber Thomson's Pacific Northwest* (Pullman: Washington State University Press, 2009), 31.

17 Thomson, *That Man Thomson*, 15.

18 David B. Williams, *Too High and Too Steep: Reshaping Seattle's Topography* (Seattle: University of Washington Press, 2015); Wilson, *Shaper of Seattle*; and Klingle, *Emerald City*, 44–85, 86–118.

19 Williams, *Too High*, 205.

20 Thrush, *Native Seattle*, 95.

21 Cronon, *Nature's Metropolis*, 74–93.

22 Wilson, *Shaper of Seattle*.

23 David Louter, *Windshield Wilderness: Cars, Roads, and Nature in Washington's National Parks* (Seattle: University of Washington Press, 2006), 3–10, 11–35.

24 Robert W. Rydell, *All the World's a Fair: Visions of Empire at American International Expositions, 1876–1916* (Chicago: University of Chicago Press, 1984), 4. See the Alaska-Yukon-Pacific Exposition section of the University of Washington Digital Collections, accessed January 17, 2020, https://content.lib.washington.edu/aypweb/index.html.

25 For examples of picture postcards, designs of the grounds, and photographs from the fair, see the Alaska-Yukon-Pacific Exposition

section of the University of Washington Digital Collections, accessed January 17, 2020, https://content.lib.washington.edu/aypweb/index.html.

26 See Erik Loomis, *Empire of Timber: Labor Unions and the Pacific Northwest Forests* (New York: Cambridge University Press, 2016); and Klingle, *Emerald City*, 160.

27 James Gregory, "Mapping American Social Movements," accessed January 17, 2020, http://depts.washington.edu/moves/about.shtml.

28 "Chinese Exclusion Act (1882)," accessed January 4, 2020, www.ourdocuments.gov/doc.php?flash=false&doc=47; and David Cahn, "The 1907 Bellingham Riots in Historical Context," accessed January 4, 2020, http://depts.washington.edu/civilr/bham_history.htm.

29 Charles Pierce LeWarne, *Utopias on Puget Sound, 1885–1915* (Seattle: University of Washington Press, 1995), 3.

30 LeWarne, *Utopias on Puget Sound*, 3–14, 168–226; and Dona Brown, *Back to the Land: The Enduring Dream of Self-Sufficiency in Modern America* (Madison: University of Wisconsin Press, 2011), 3–17.

31 Colin M. Anderson, "The Industrial Workers of the World in Seattle General Strike," accessed January 4, 2020, https://depts.washington.edu/labhist/strike/anderson.shtml.

32 William G. Robbins and Katrine Barber, *Nature's Northwest: The North Pacific Slope in the Twentieth Century* (Tucson: University of Arizona Press, 2011), 127–90.

33 Lewis Mumford, for the Northwest Regional Council, *Regional Planning in the Pacific Northwest: A Memorandum* (Portland: Northwest Regional Council, 1939), 15.

34 Mumford, *Regional Planning in the Pacific Northwest*, 20.

35 White, *Organic Machine*, 64; and Mumford, *Regional Planning*, 20.

36 David Wilma, "President Roosevelt Approves Loan for Yesler Terrace Public Housing Project in Seattle on December 2, 1939," HistoryLink.org, January 1, 1999, www.historylink.org/File/2105; and Megan Asaka, "'40-Acre Smudge': Race and Erasure in Prewar Seattle," *Pacific Historical Review* 87, no. 2 (2018): 231–63.

37 See Adam Rome, *The Bulldozer in the Countryside: Suburban Sprawl and the Rise of American Environmentalism* (New York: Cambridge University Press, 2001); and Klingle, *Emerald City*, 203–29.

38 Findlay, *Magic Lands*, 214–64; and David Farber, *The Age of Great Dreams: America in the 1960s* (New York: Hill and Wang, 1994).

39 Richard C. Berner, *Seattle Transformed: World War II to Cold War* (Seattle: Charles Press, 1999); and Sanders, *Seattle and the Roots*.

40 Klingle, *Emerald City*, 203–29.

41 Sanders, *Seattle and the Roots*, 131–214; and Andrew G. Kirk, *Counterculture Green: The* Whole Earth Catalog *and American Environmentalism* (Lawrence: University Press of Kansas, 2007), 43–155.

42 Callenbach, *Ecotopia*, 98; and Sanders, *Seattle and the Roots*, 65–98. Writing about the region he termed *Ecotopia* in the 1980s, journalist Joel Garreau described a Northwest coming into view: "In Ecotopia even the middle class has moved on the idea that a person may have to lower [their] . . . standard of living in order to raise [their] overall quality of life." Joel Garreau, *The Nine Nations of North America* (Boston: Houghton Mifflin, 1981), 262.

43 Binda Colebrook, *Winter Gardening in the Maritime Northwest: Cool Season Crops for the Year-Round Gardener*, new rev. ed. (Seattle: Sasquatch Books, 1989).

44 Sanders, *Seattle and the Roots*, 65–98; Michael Pollan, *The Omnivore's Dilemma: A Natural History of Four Meals* (New York: Penguin, 2006); and "P-Patch Community Gardening," Seattle Department of Neighborhoods, accessed January 17, 2020, www.seattle.gov/neighborhoods/programs-and-services/p-patch-community-gardening.

45 Judy Mattivi Morley, *Historic Preservation and the Imagined West: Albuquerque, Denver, and Seattle* (Lawrence: University Press of Kansas, 2006), 124–26; and Sanders, *Seattle and the Roots*, 16–64.

46 Klingle, *Emerald City*, 234–35.

47 For a discussion of this era and the rise of Microsoft, see Fred Moody, *Seattle and the Demons of Ambition: From Boom to Bust in the Number One City of the Future* (New York: St. Martin's Griffin, 2003).

48 Richard Florida, *Cities and the Creative Class* (New York: Routledge, 2005); and Andrew Marantz, *Antisocial: Online Extremists, Techno-utopians, and the Hijacking of the American Conversation* (New York: Viking, 2019).

49 *Seattle/King County Point-in-Time Count of Persons Experiencing Homelessness, 2017*, accessed March 25, 2020, http://allhomekc.org/wp-content/uploads/2016/11/2017-Count-Us-In-PIT-Comprehensive-Report.pdf; and David Kroman, "Entering the Third Year of Seattle's Homelessness Emergency," Crosscut, December 28, 2017, https://crosscut.com/2017/12/seattle-homelessness-emergency-jenny-durkan.

50 "The Spheres," accessed February 21, 2020, www.seattlespheres.com.

51 Daniel Beekman, "Tech Giant's Seattle Campus a Backdrop for 'Tax Amazon' Rally," *Seattle Times*, updated April 12, 2018.

52 Forward Thrust Committee, *Developing a Capital Improvement Plan for King County, Part III: Recommendations* (Seattle, 1967); and Alyssa Burrows, "King County Voters Approve Proposition 6, a Forward Thrust Parks and Recreation Bond, on February 13, 1968," HistoryLink.org, January 3, 2003, https://historylink.org/File/4116.

53 "The Spheres"; and Jon Talton, "Vote: Will Amazon's Spheres Replace the Space Needle as the New Symbol of Seattle?," *Seattle Times*, January 26, 2018.

CHAPTER 2

THE MAKING OF URBAN CASCADIA

Extending Urbanization through Airplanes, Software, and Infrastructure

NIK JANOS

DRIVING NORTH ON US HIGHWAY 101 FROM ASTORIA, OREGON, you cross the Columbia River, fourteen miles from the mouth, on the Astoria-Megler Bridge. It is the longest truss bridge in the United States, about four miles across. The river is wide and powerful, and just looking at it you cannot see the extraordinary transformations people have carried out over the past one hundred plus years. From the mouth of the river up multiple hundreds of miles, from the nineteenth century through the twentieth century, capitalists and government officials constructed an "organic machine" from one of North America's largest rivers.[1] They built fourteen dams that enable vast amounts of hydropower, which supplies a majority of electricity to the region. The dams supply water to the otherwise dry eastern side of the Cascade Mountains, turning it into a hugely productive agricultural region. In earlier eras, the river was an aquatic highway for moving goods and lumber. Until the twentieth century, before the toll of the dams and overfishing, the river was an ecosystem abundant with salmon. Standing next to the Grand Coulee Dam, the largest hydropower station in the United States, picking a Washington apple, or sipping coffee in downtown Portland are all degrees of the new socionature that the organic machine has enabled. The Columbia River is but one example of the urban-industrial metabolism that weaves Cascadia into an urban fabric.[2] In this chapter, I argue that to understand urban Cascadia we must look beyond the cities of Vancouver, Seattle, and Portland to the vast and interconnected transportation, energy, and water infrastructure and the emerging political imaginary that are connecting the region in coherent yet uneven ways.

In advancing this argument, this chapter takes a relational and regional approach to the urban environmental history of Cascadia. This approach starts with recognizing that the distinction between the categories of "country" and "city," so prevalent in the nineteenth and twentieth centuries, no longer make sense in light of urban scholarship that shows how these seemingly different realms are in fact mutually constituted by capitalist urbanization.[3] Replacing this worn dichotomy is the generative understanding of urbanization as a hybrid process that melds nature and society—socionature—into uneven geographies via capitalist commodity flows and infrastructure, state policy, law, and investments, and humans and nonhumans, who participate as well as resist.[4] The infrastructures of extended urbanization have "invested metropolitan actors in the transformation of distant landscapes while drawing distant people into new relationships with metropolitan centers."[5] Urbanization, therefore, intensifies and extends the relationship between nature and humans at multiple scales, from local to regional to planetary (see introduction).[6] Consequently, this chapter considers how urban and industrial development in the region contributes to and is shaped by planetary urbanization and capitalism and by political structures and institutions far afield. Rather than retell the city-centric story of Vancouver, Seattle, and Portland, this is a story of urbanization as a process that knits nature, space, place, and territory together from a regional perspective.[7]

I begin the chapter by considering how three regionally grown corporations, Boeing, Microsoft, and Amazon, are shaping capitalist urbanization in Cascadia and around the planet. The bulk of the chapter is then dedicated to exploring urban Cascadia through the lens of transportation, energy, and water infrastructure and the governance practices that facilitate their development. I show that regional integration, both materially and politically, has been developing for the better part of a century. It is only at the end that I examine the creation of the three big cities of Cascadia and the role that city and suburb making plays in stitching together urban Cascadia.

The examples I provide in this chapter are in no way exhaustive but are representational. It is crucial to note that the true beginning of this story is with the Indigenous peoples of the region and their fate in the face of settler colonialism.[8] That story, however, will be taken up by Erin Goodling, Coll Thrush, and Danica Miller later in this book.

EXTENDED URBANIZATION, CASCADIA STYLE: AIRPLANES, SOFTWARE, AND THE INTERNET

Around 4:00 a.m. on August 9, 1945, a Boeing B-29 plane, designed and mostly assembled in Renton, Washington, and tested at Boeing Field in Seattle, took off from a small island in the Pacific Ocean and headed toward mainland Japan. At 11:02 a.m., Nagasaki time, the second atomic bomb ever dropped as an act of war, containing plutonium enriched at Hanford nuclear facility (see map I.1 in introduction) on the banks of the Columbia River in Washington, exploded over Nagasaki, instantly killing between 40,000 to 75,000 people. Washington-based Boeing Corporation played an integral role in both theaters of World War II, their planes carrying the payload for massive death tolls, urban incineration, and environmental impacts that ultimately helped the United States to become victorious over Germany and Japan. Since 1945, Boeing airplanes, made possible by massive federal spending, have accelerated a double movement, one that bolsters US global hegemony as well as funds industrial and urban growth throughout the region.

Bill Boeing, the son of a timber magnate, moved to Hoquiam, Washington, in 1903 at twenty-two years of age, looking to start his own timber business.[9] Using the capital generated from felling the trees of Washington, and after a chance encounter with a budding aviation enthusiast, Boeing launched his aircraft startup in 1913.[10] Although the corporate headquarters is now in Chicago, design, testing, and assembly are still carried out in Renton and Seattle, making the region the epicenter of the global aviation revolution of the twentieth century. Boeing is the world's largest civilian aircraft manufacturer, the European-made Airbus only right behind. Boeing has helped transform the geography of twentieth- and twenty-first-century globalization, leading to the key dimension of the compression of space and time, enabling people to move around the planet with relative ease.

In 1998, Redmond, Washington–based Microsoft Corporation was at its pinnacle of power. It controlled 86 percent of all personal computer operating systems, the only competition being—at the time—a dying bit player named Apple Computer. Microsoft was successfully implementing its "Windows everywhere" strategy, aimed at getting original equipment manufacturer (OEM)–built computers running Windows OS on every desktop in the world. Though a thin stretch of the San Francisco Bay Area, known as Silicon Valley, was the original center of computer hardware and software innovation, by the late 1990s, Redmond was accumulating most of the revenue and profits on the software side, as hardware manufacturers raced to the bottom on thin margins. Although Microsoft missed the rise of the smartphone, led ironically by Apple, Microsoft not only changed Redmond but also has had a profound impact on myriad facets of modern life, from personal computing, academics, database management, corporate productivity, and more recently cloud computing with the Microsoft Azure platform. Like Boeing, Washington-based Microsoft—and Bill Gates—has played a part in transforming large parts of daily life around the world.

Whether one is on a bus or lying in bed at 2 a.m., on any smartphone one can tap the Amazon.com app, find that book or pair of shoes, and tap "Buy Now," Amazon's patented one-click buying button. Moments later a notification arrives indicating that the item will be delivered tomorrow, one-day shipping. Soon after, in a giant warehouse guided by proprietary logistics and inventory algorithms,[11] a low-paid and overworked human starts to fill the order, one of thousands she will fulfill that day. This sequence is repeated each day by millions of people across increasingly large swaths of the planet, which has reshaped the production, distribution, and consumption of commodities. With its logistics and industry-leading Amazon Web Services, Seattle-based Amazon has built vast new assemblages linking Western brands and OEMs in China with the proliferation of a constellation of Amazon distribution centers and online sales channels in the some of the wealthiest markets in the world, most notably the United States. Like strip malls and big box stores before, online retail, along with the physical infrastructure of shipping, warehousing, delivery, consumption, and waste, shape the processes of sub/urbanization and the compression of space and time. Like a number of fabled tech companies, Jeff Bezos started Amazon in his modest suburban house selling books online. In 2017, he surpassed Bill Gates as the richest man in the world. Amazon, Apple, Microsoft, Google, and Samsung are regarded as members of the five horsemen of technology,[12] all competing and intertwining to dominate the internet technologies of the twenty-first century. Although not typically considered part of urban history, the development of airplanes, software, and the internet is an illustrative example of how the process of extended urbanization in Cascadia knits space, place, and territory together within global history.

URBANIZATION THROUGH TRANSPORTATION INFRASTRUCTURE

PORTS

Before the internet, planes, trains, and trucks, the land taken from the Indigenous people of Cascadia was first connected to colonial powers and economies by sea. The first Europeans to reach the region were Spanish and English explorers in the 1770s. Like all colonial expeditions, early explorers were interested in one of three things: trade, military outposts, or land. It was not until Lewis and Clark entered the region in 1803, the first people of European descent to explore the region by land, that the United States entered the colonial territorial geopolitics of northwestern North America. However, because it was easier to reach these territories by sea, ports were instrumental in incorporating the region into global colonial politics as well as national and international trade networks. When the industrialists arrived in the middle of the nineteenth century, ports took on a greater significance as industrial capitalists needed to find export markets for the commodities churning out of the region's lumber mills and salmon canning factories. It was in fact lumber that elevated the region's importance and made the port cities of Vancouver and Seattle, and to a lesser extent Portland, vital hubs of globalizing commodity networks via the Pacific Ocean. The lumber industry of the region reached dominance between the late 1800s and World War II, aided in part by the seaports as well as the intercontinental railroads.

The port cities of Cascadia have played a vital role in regional, national, and global economies and politics. For example, in the late

1800s, the Oregon Steam Navigation Company came to dominate human and commodity transport on the Columbia River, which stitched together Portland, the Willamette Valley, and the territories of Eastern Oregon, Eastern Washington, and Idaho, in the process becoming Portland's first million-dollar company.[13] During World War I, the port of Portland played a significant role producing nearly a hundred steel-hull and wooden-hull ships for the US military in its counterstrategy against Germany's U-boat campaigns.[14] Later in the twentieth century, the region was locked into a Pacific Ocean lumber triangle trade between British Columbia, Washington/Oregon, and East Asia (Japan and China).[15] However, the timber and lumber industries began their decline after the 1980s and have continued to decline ever since.[16]

Although the largest ports on the West Coast of North America are in California, Vancouver and Seattle are the major conduits of global flows of commodities in and through Cascadia. The ports are no longer dominated by lumber. Now they facilitate—along with Amazon, Microsoft, Apple, and the like—the inflow of electronic commodities and manufactured goods from East Asia and the outflow of agricultural, fish, and fossil fuel commodities from Cascadia to Asia. When you think of Seattle-based REI, think of China as much as hiking in the Cascade Mountains.

ROADS AND HIGHWAYS

The Columbia River was the nineteenth- and early-twentieth-century aquatic highway, which was joined by and then overtaken by the intercontinental railroads. Today, the immense networked interstates, international highways, roads, streets, and dirt roads enable the flow of goods, services, and people to just about every nook and cranny of Cascadia and beyond. This is not something that is unique to the region. Similar to the rest of the car-centric North America, what people call urbanization is stitched together by this vast network of mostly publicly funded road infrastructure. Washington and Oregon are connected to the rest of the United States by three major interstates: I-90 and I-84 running east-west and I-5 running north-south. These major highways were created by the federal government in 1956 with the passage of the Federal Aid Highway Act of 1956 in what was one of the country's largest public works projects.[17] There are also lesser interstate highways that preceded the aforementioned major highways. These include Highway 101, Highway 395, Highway 20, and Highway 95. British Columbia is connected to the rest of Canada via Highway 1 (trans-Canada highway) and Highway 16, whereas Highway 93 and Highway 99, among others, connect British Columbia to the United States. The latter, Highway 99/I-5, is the symbolic and material transportation infrastructure that connects Cascadia together.

MAINSTREET CASCADIA: A MEGAREGION IN THE MAKING

In another example of how infrastructure plays an integral role in building urban Cascadia as well as place branding, Highway 99/I-5, which runs south from British Columbia, west of the Cascade Mountains, and on through California to Mexico, has become a centerpiece of a recognized urban megaregion (see map I.1). Highway 99/I-5 is 1,350 miles long with about 550 miles in heavily congested urban areas. On any given day, between 71,000 and 300,000 vehicles travel on I-5 alone, with about 10,000 to 35,000 commercial trucks daily.[18] As a piece of transportation infrastructure, Highway 99/I-5 is certainly critical to the movement of people and commodities up and down the West Coast. However, it is the portion from Vancouver, British Columbia, to Eugene, Oregon, that transcends its role as transportation infrastructure. As discussed in this volume's introduction, Mainstreet Cascadia is a way to describe the space that is the thin strip of land between the Cascade Mountains and Salish Sea, the Pacific Ocean, and the Coastal Range in Oregon (see map I.1). This socionature is composed of the largest concentration of inhabitants in the region—over ten million people—built environments, cities, ports, and businesses. The Mainstreet Cascadia metaphor allows us to visualize the interlinked economic, social, political, and ecological forces, structures, and networks that form a functional and ideological sense of place in Cascadia. It is the clearest sense of an urban socionature, or urban fabric, of this portion of the North American continent. It is the binational, West Coast version of the Boston-Washington megaregion of the eastern United States.

RAIL

The proposed high-speed rail network that would connect Cascadia from British Columbia to Oregon is the latest manifestation of urbanization through infrastructure in Cascadia.[19] The region is already connected by the aptly named Amtrak Cascades, which connects Eugene, Oregon, and Vancouver, British Columbia. Though it is listed as a semi-high-speed network, travel times are on par with or exceed automobile transport on Highway 99/I-5. For example, Seattle to Vancouver on Amtrak Cascades is 4 hours and 30 minutes, yet by car, it is only 2 hours and 41 minutes.[20] The Portland to Seattle time is quicker at 3 hours and 30 minutes on Cascades and 3 hours and 14 minutes by car.[21] Ridership on Cascades saw tremendous growth from the initial link between Portland and Seattle at 94,000 annual riders in 1993, increasing to 287,000 people the year that the link to Vancouver was opened, and in 2011, the system reached its historic high at 848,000 riders.[22] The system averaged about 805,000 riders per year between 2009 and 2018.[23]

However, because the existing service is relatively slow, political and economic leaders see it as an impediment to the closer integration of the urban megaregion.[24] Consequently, the province of British Columbia, the state of Washington, and the state of Oregon, with the Microsoft Corporation playing a large private role, are coordinating the new high-speed rail project. According to the official business case analysis,[25] the high-speed rail network would connect between 1.7 million to 3.1 million people with an operating revenue ranging from $160 million to $250 million the first year. At its fastest, the trains would travel 217 mph and cut the times from Vancouver to Seattle and Seattle to Portland to an hour. The study reports that construction could begin in 2027 and be completed in 2034, with the total cost between $24 billion to $42 billion.[26] Ultimately, planners want the high-speed rail network to be cost and time competitive with air and car travel.

The high-speed rail project is the paragon of infrastructure mega projects that aims to integrate the economies, political institutions, territories, peoples, and space of the region. In this sense, the proposed high-speed rail is also a grand attempt to build Cascadia as a place and political imaginary, albeit an eco-capitalist one. Washington State and its environmentalist governor Jay Inslee are taking a leading role in crafting the vision and branding of Cascadia via the high-speed rail project. Inslee writes, "My vision for the megaregion—stretching from Washington, north to British Columbia, and south to Oregon—includes a transportation system that is fast, frequent, reliable, and environmentally responsible. Such a system would unite us in our common goals related to economic development, shared resources, affordable housing, new jobs, tourism, multimodal connections, and increased collaboration."[27] Proponents herald this project for its transportation, environmental, and climate credentials, but additionally they argue that it "unites this Cascadia megaregion, and positions it for global competitiveness and future prosperity."[28] The high-speed rail system was the linchpin of the memorandum of understanding that Governor Inslee and British Columbia premier Christy Clark signed at the first Cascadia Innovation Corridor conference in 2016. Proponents argue that the train would enable the growth of a new technology and innovation corridor, connecting high-tech workers to new and existing high-tech corporations along Mainstreet Cascadia. All told, the high-speed rail project represents a binational, multiscalar, private-public, cultural, economic, and political endeavor to stitch the actually existing urbanizing megaregion together with physical infrastructure and the territorial imaginaries of Cascadia. So far, it is the clearest manifestation of governing through infrastructure of the political territorial imaginary called Cascadia.

THE ENERGY LANDSCAPES OF URBAN CASCADIA

ENERGY PRODUCTION

Energy and power systems, like other large infrastructure, are administered and territorialized within levels of governance, from national, state, and municipal. Moreover, energy and electricity systems are mixtures of private, public, and public-private ownership. Adding more complexity, people account for energy and power systems in the realms of production and consumption with a plethora of institutions, laws, and organizations managing these flows all along the "pipeline." Furthermore, accounting is broken down between energy and electricity, with the former consisting of primary energy for industrial purposes and heating systems and the latter most familiar as a switch or outlet on a wall. All told, energy and electricity systems are one of the most complex political, economic, and social systems of modernity. Although energy is produced and sold as a commodity in its own right, the larger significance for this chapter is how people put energy and electricity to use in fueling urbanization.

The Cascade Range and the Canadian Rockies and the abundant rain and snow that produce the fast-flowing rivers have made Cascadia an ideal location for hydroelectric power generation. The Columbia River is the par excellence hydroelectric river in the region for both the United States and Canada. Additionally, there are numerous other rivers that have been dammed to provide electricity as well as water for end-use consumption. Washington, Oregon, and British Columbia all operate electric grids that are dominated by hydroelectric power. In Washington, over two-thirds of the state's electricity is hydroelectric, and in British Columbia, hydroelectricity is 90 percent of the electricity.[29] In Oregon, hydroelectricity is about 50 percent of electricity; however, it can range from 70 percent to 90 percent in extremely wet years.[30] The Grand Coulee Dam on the Columbia River dominates hydroelectric production, supplying power to eleven western states and at full capacity can power 2.3 million households.[31] The Columbia more broadly contributes the lion's share of hydroelectric power for the three subnational territories of Cascadia. For example, Oregon's top four electricity-generating facilities are hydroelectric dams on the Columbia River (John Day, The Dalles, Bonneville, and McNary).[32] Due to the Columbia River Treaty of 1961, BC Hydro, the provincial hydroelectric company, is entitled to half of the hydroelectricity produced on the US portion of the river. British Columbia has a new large-scale hydroelectric dam under construction on the Peace River. The project, called Site C, will come online in 2024 and will provide 1,100 megawatts of capacity, enough to power about 450,000 homes per year.[33]

Although hydroelectricity dominates the electric grids of Washington, Oregon, and British Columbia, they rely on other energy sources to fill gaps in the variability of hydro as well as non-hydro energy for industrial activity. In Oregon, natural gas is used to fill in the gaps and at times is about 20 percent to 25 percent of the electricity mix, depending on how much hydroelectricity is produced. Wind power, from nearly 1,900 turbines, contributes 12 percent of electricity.[34] At the margins are coal, which is only ever about 3 percent of Oregon's electricity mix, biomass, and solar.[35] In Washington, the electricity mix is 80 percent hydro, but the overall mix of renewables can climb as high as 90 percent when adding in biofuels and thermal energy from woodburning stoves to heat homes. The remainder of the baseline electricity in Washington comes from its sole nuclear power plant on the Columbia River; a coal-fired plant in Centralia, which is set to be retired by 2024; and natural gas plants, which will pick up the slack after the coal plant closes. Wind energy is growing in the state and had a slight boost back in 2001 when the state's first utility-scale project came online. Currently, wind is the fourth largest electricity generator. In British Columbia, biomass from the region's lumber industry contributes about 5 percent, and natural gas makes up the remaining baseline source. Throughout the region, the electric grids are overwhelmingly non-fossil fuel with hydro dominating the mix. However, natural gas, biomass from the lumber industry, and one nuclear power plant provide the baseline energy required in a grid with a high dependence on variable hydro. Wind and solar are at present miniscule.

ENERGY CONSUMPTION

Energy and electricity production are only one aspect of a given geography's or territory's energy story. How electricity and energy are used and consumed are equally important and reveal a different story. End use reveals how the region is deeply connected to national and international flows of fossil fuels to power everything that is not electrified. As in most places in North America, in Cascadia, transportation is the largest end-use sector, and this means that petroleum is the largest non-electricity end-use energy source. In terms of end-use sectoral categories in Washington, transportation is 32.4 percent of end-use consumption, and industrial and residential use are tied at about 25 percent each, and commercial use rounds out at 18 percent.[36] Oregon is nearly identical with industrial and residential use both around 25 percent, transportation at 29.5 percent, and commercial use at 19.6 percent.[37] In British Columbia, things are quite a bit different. Industry accounts for 46 percent of energy and electricity use, and transportation accounts for 29 percent. Residential use and commercial use are much lower than Washington and Oregon at 13 percent and 12 percent, respectively.[38]

As discussed above, residential end use and commercial end use are primarily supplied by hydroelectric power with fossil fuels and nuclear providing baseline energy. A small percentage of renewables such as utility-scale wind and biomass and rooftop solar round out the mix. Transportation and industry are fossil fuel intensive end-use sectors. Washington and Oregon are net importers of petroleum and natural gas, having almost no fossil fuel production. Washington, for example, imports oil from Alaska and more recently from the Bakken Shale Formation in North Dakota. Petroleum is refined at large refineries in Washington, which are the fifth largest in capacity in the United States. Washington has recently increased its importation of natural gas, most of it coming from British Columbia. However, about two-thirds of the natural gas that enters Washington continues on to Oregon and California. Oregon relies on petroleum refined in Washington for nine-tenths of all its petroleum use. British Columbia, in contrast, has both crude oil production and natural gas production, which means end users consume some of these resources. British Columbia produces about 75,500 barrels per day of crude oil, which make up about 1.7 percent of Canadian production. Oil is refined in one of two refineries in Prince George and Parkland in Burnaby. British Columbia has 29 percent of Canadian natural gas production, combined natural gas and liquified natural gas. Natural gas is produced in the northeastern part of British Columbia at the Montney formation. Recent development of "tight gas" led to large increases in natural gas production between 2006 and 2017.[39]

ENERGY INFRASTRUCTURE AND GOVERNANCE

Energy landscapes are made by more than production and consumption. Energy infrastructure to transmit and govern energy systems ties production and consumption together. Indeed, "electricity does not so much flow, as we are accustomed to thinking about the motion of commodities in a market system, as flash, annihilating the space between consumption and production."[40] Transmission and governance also give us the clearest picture of how the political territories and governments, public and private energy companies, and residents of Cascadia are entwined together within an energy nexus. Additionally, it exemplifies how the state funded expensive energy infrastructure but inexpensive hydroelectricity and power fuel urbanization, which allows city building, industrial development, and agriculture development across the region.

The governance of the Cascadia energy nexus is a web of private, public, and private-public organization and city, state, and federal governments. The aforementioned Columbia River Treaty is a good example of binational governance of flood risk and electricity production, transmission, and consumption. The original impetus for the treaty was in flood control, pushed mostly by the US government to protect population centers and agricultural interests. As mentioned above, the treaty guaranteed British Columbia half of the electricity generated on the US side of the Columbia River. The treaty is administered by the two nation-states. On the US side, the treaty is administered by the Bonneville Power Administration (chair) and the Northwestern Division Engineer (member) of the US Army Corps of Engineers. Both of these are federal agencies. On the Canadian side, the treaty is administered by British Columbia's publicly owned Hydro and Power Authority (BC Hydro).[41]

One further example is illustrative of the ways that energy infrastructure and governance tie the region together. The Northwest Power Pool is a corporation based in Portland, Oregon, that helps coordinate electrical utilities that own power-generating plants to sell power throughout the US Pacific Northwest and Western Canada. Most of the power-producing utilities, as discussed above, are hydroelectric power-generating businesses, and the Power Pool assists with power planning in the region by setting "critical reservoir elevation limits (rule curves) for US dams and . . . contingency reserve power requirements for utilities within its geographic area."[42] Although the Power Pool is a voluntary organization, it has emerged as an important coordinating body to facilitate energy generation, distribution, and price controls and the ways that energy producers integrate throughout the region, which also impacts their ability to integrate into the high-voltage transmission networks that send electricity to utilities and ultimately consumers as far off as California and Mexico and east to Montana.[43]

Moves to integrate the region's utilities actually began in the 1920s when the municipal electric systems of Tacoma and Seattle were interconnected. In the intervening years, more and more local and regional utilities were integrated. When the Grand Coulee Dam became operational, this opened another opportunity to scale up the region's electricity integration. In the early 1940s, the US federal government built high-voltage transmission lines to load centers west of the Cascade Mountains. Further expansion occurred in 1956, when the federal Bonneville Power Administration interconnected with Idaho Power on the border of Oregon and Idaho. The final piece of the regional interconnection was the creation of the Northwest Power

Pool itself during World War II. The Power Pool was created when the federal government mandated that electric utilities pool their operations in order to optimize for wartime mobilization. In the 1960s, the Power Pool was a critical component of implementing the 1964 Pacific Northwest Coordination Agreement, which brought coordinating efforts of electricity generators on the Columbia River in alignment with the newly signed Columbia River Treaty. This alignment scaled the Power Pool's reach into binational as well as regional electricity production, transmission, and market.[44]

WATER FOR AGRICULTURE AND URBANIZATION

In Cascadia, water is not just for electricity; it is also the cornerstone of the region's agriculture sector and provides a diverse set of uses for the region's fifteen million inhabitants. Therefore, the region's dams do more than generate electricity; they generate the capacity to grow food and grow cities, suburbs, and exurbs (see chapter 4), among other things. Agriculture in the region is concentrated by sector in particular biomes of the state, depending on soil quality and water availability. For example, in Washington, most of the agriculture is concentrated east of the Cascade Mountains and relies on the Columbia River; in Oregon, east of the Cascades is mostly ranch land, and west of the Cascades, various microclimates are good for nurseries, fruit trees, and dairy farms. Washington has 35,700 farms of various sizes operating on a total of 14.7 million acres.[45] Oregon has 37,200 farms operating on a total of 16 million acres.[46] In contrast, British Columbia has just 17,528 farms on just 6.1 million acres (2.5 million hectares), mostly located in the Fraser Valley.[47]

Although California in the United States and Saskatchewan in Canada are North America's "breadbaskets," Washington, Oregon, and British Columbia are leading producers of key agricultural commodities, consumed in and exported to regional, national, and global markets. Of the three, Washington has the largest agriculture sector by value, coming in at $10.6 billion in 2017. The most famous agricultural commodity from Washington is the apple, which is a $2.43 billion a year industry, accounting for 23 percent of the state's total agricultural value.[48] Seventy percent of all apples consumed in the United States are grown in Washington. Similarly, if you enjoy beer, then chances are great that the hops in your beer were grown in Washington. Other commodities that Washington is a number one producer of in the United States are spearmint oil, wrinkled seed peas, and concord grapes.[49] While Oregon has more operations and acres devoted to agriculture, its agriculture sector is considerably less valuable, coming in at $4.9 billion in 2019. Interestingly, Oregon's main valued agricultural commodities are seeds, oilseeds, and other horticultural products. In particular, Oregon's greenhouses and nurseries produce almost one billion dollars per year, followed by hay ($674 million), cattle and calves ($642 million), milk ($552 million), grass seed ($517 million), and wheat ($282 million).[50] British Columbia is not a major agriculture center by acreage; however, food exports were about $2.6 billion in 2017, up from $1.6 billion in 2008, most of which (72 percent) were exported to the United States.[51] Fish and seafood, and in particular farmed Atlantic salmon, are British Columbia's principal food export, with health food products and baked goods and cereals coming in second. All told, water and energy infrastructure and its governance are key aspects of how extended urbanization stitches Cascadia together. The energy landscapes of Cascadia constitute the networks and nodes as well as the metabolic flows that enable city and suburban territories to proliferate throughout the region and beyond.

THE SPATIAL IMPLOSION AND EXPLOSION OF URBAN AND SUBURBAN CASCADIA

Lastly, I will briefly consider how the city territories and sub/urban agglomerations fit into the history of urban Cascadia (for more detailed accounts, see chapters 1, 5, 7, and 8). Seattle, Vancouver, and Portland share many similarities in their fabled, humble origins. Indigenous people had been living there since time immemorial when the city builders, the "heroic great men," that is, the land-thieving white men, arrived and dispossessed them of the land in the late nineteenth century. This, of course, belies the full set of diverse people that built these cities (see chapter 5). In the history of all three cities, three conjunctures explain their eventual regional domination. First, abundant raw materials, particularly trees, drew settler colonists from around the world to the region, but regional urbanization was jump-started when the aforementioned city builders arrived, turning capital accumulation from resources into urban development. Second, because these cities were some of the last major cities built in North America, at their time of origin, they were built into a preexisting and globalizing capitalist world economy and European colonial geopolitical world order, which facilitated rapid spatial and economic expansion. Third, despite the strength of the Pacific Rim economy, it was not until the intercontinental railroads connected these cities to their national territories, and thus to greater influx of capital, political control, and importantly to hundreds of thousands, and eventually millions, of settlers, that these cities emerged as regional and ultimately national power centers.

In the late twentieth and early twenty-first centuries, the processes of sub/urbanization, industrialization, and globalization emanating from within the region and shaping it from beyond contributed to massive urban-environmental changes in Cascadia. Dams provided inexpensive energy and water that allowed the region's cities to grow upward and outward and consequently altered the ecological conditions throughout the region. This process of low-density and car-centric suburbanization, also made possible by publicly funded roads and highways, accelerated after World War II, making it the dominant process and form of urban growth in the region. The City of Seattle, for example, embarked on an aggressive territorial expansion between the 1950s and 1970s via the process of annexation. In 1970, the City explained that "without the annexation area, the city would be about 100,000 persons smaller than it is. It would be less well represented in the state legislature, would include proportionally fewer single-family homes, proportionally more older structures, have proportionally less commercial tax base, would be proportionally more dense [*sic*], have proportionally less vacant land for new development, and would not [have] had the participation in public affairs of many able people."[52] On this point, Roger Keil argues that the urban century is really the suburban century: "most activity in terms of the expansion and contraction of urban population, built form and economic activity will occur in peripheral areas."[53] Urban Cascadia is located in the periphery as much as the central city, in suburbs such as Surrey, British Columbia, Bellevue and Redmond, Washington, or Beaverton, Oregon, or further afield in exurbs such as Hood River and Bend, Oregon (see chapter 4). Only in the 1990s and later did elites grab hold of green urbanism and begin to focus on density in the core of these territorial cities (see chapters 8–10).

CONCLUSION

In this chapter, I have endeavored to tell a relational story of the urbanization of Cascadia, attempting to avoid a city-centric focus. Spatial history unfolds via the simultaneous process of implosion and explosion, to use Henri Lefebvre's words.[54] Urbanization and industrialization exploded across the region, connecting more and more territories, spaces, biomes, and humans and nonhumans, while also imploding through the concentration of political and economic power, human labor, and resources in increasingly concentrated nodes throughout the region. These are not distinct stories, and in this chapter, I examined their relationships through networks, multiscalar territorial governance, commodity flows, and interlinked built environments. The story I told was a deeply relational story that attempts to map and describe the multiscalar, place-based, and territorial relationships that have been built across the region but critically operate in other places, territories, and scales. From atomic bombs dropped from Boeing airplanes, to Microsoft Word with which I type these words, to the plethora of Amazon delivery trucks rumbling down suburban streets, to deep urban and rural inequalities that form the hidden fissures of the green utopia, to Fuji apples sold in stores in New Orleans, Louisiana, to flipping on the light switch at the brewpub in Eugene, Oregon, which by the way just sold to a transnational conglomerate, the implosions and explosions of urban Cascadia have and will continue to reshape the region and beyond.

NOTES

1 Richard White, *The Organic Machine: The Remaking of the Columbia River* (New York: Hill and Wang, 1995).

2 Joshua P. Newell and Joshua J. Cousins, "The Boundaries of Urban Metabolism: Towards a Political-Industrial Ecology," *Progress in Human Geography* 39, no. 6 (December 2015): 702–28, https://doi.org/10.1177/0309132514558442.

3 William Cronon, *Nature's Metropolis: Chicago and the Great West* (New York: W. W. Norton, 1991); and Matthew Gandy, *Concrete and Clay: Reworking Nature in New York City* (Cambridge, MA: MIT Press, 2002).

4 Nik Heynen, Maria Kaika, and Erik Swyngedouw, eds., *In the Nature of Cities: Urban Political Ecology and the Politics of Urban Metabolism* (London: Routledge, 2006).

5 Andrew Needham, *Power Lines: Phoenix and the Making of the Modern Southwest* (Princeton, NJ: Princeton University Press, 2014), 10.

6 Yannis Tzaninis, Tait Mandler, Maria Kaika, and Roger Keil, "Moving Urban Political Ecology beyond the 'Urbanization of Nature,'" *Progress in Human Geography* (February 2020), https://doi.org/10.1177/0309132520903350.

7 For excellent city-specific books on these cities, see Carl Abbott, *Portland in Three Centuries: The Place and the People* (Corvallis: Oregon State University Press, 2011); Matthew Klingle, *Emerald City: An Environmental History of Seattle* (New Haven, CT: Yale University Press, 2007); and Robert A. J. McDonald, *Making Vancouver: Class, Status, and Social Boundaries, 1863–1913* (Vancouver: UBC Press, 1996).

8 For outstanding accounts of Indigenous history of the region, see Penelope Edmonds, *Urbanizing Frontiers: Indigenous Peoples and Settlers in 19th-Century Pacific Rim Cities* (Vancouver: UBC Press, 2010); Alexandra Harmon, *Indians in the Making: Ethnic Relations and Indian Identities around Puget Sound* (Berkeley: University of California Press, 2000); and Coll Thrush, *Native Seattle: Histories from the Crossing-Over Place* (Seattle: University of Washington Press, 2007).

9 Robert J. Sterling, *Legend & Legacy: The Story of Boeing and Its People* (New York: Saint Martin's Press, 1991), 1.

10 Sterling, *Legend & Legacy*, 2.

11 Miriam Posner, "See No Evil," *Logic*, no. 4 (April 1, 2018), https://logicmag.io/scale/see-no-evil/.

12 M. G. Siegler, "The Fifth Horseman: Samsung," TechCrunch, January 5, 2013, https://techcrunch.com/2013/01/05/the-fifth-horsemen-of-tech-samsung/.

13 Abbott, *Portland in Three Centuries*, 36.

14 Abbott, *Portland in Three Centuries*, 97.

15 Jean M. Daniels, "The Rise and Fall of the Pacific Northwest Log Export Market," United States Department of Agriculture, February 14, 2005, www.fs.fed.us/pnw/pubs/pnw_gtr624.pdf.

16 Daniels, "Rise and Fall."

17 US Department of Transportation, "Highway History," accessed October 21, 2019, www.fhwa.dot.gov/interstate/interstatemyths.cfm.

18 Washington State Department of Transportation, "About West Coast Green Highway," accessed October 21, 2019, www.westcoastgreenhighway.com/about.htm.

19 The official terminology of the project is *ultra-high-speed*, but this reflects that the existing Cascadia line is classified as *semi-high-speed*. I am using the standard terminology *high-speed rail* because this project fits within existing high-speed rail parameters seen across the world.

20 WSP, *Ultra-High-Speed Ground Transportation Business Case Analysis: Final Report*, Appendix D, July 10, 2019, 6.

21 WSP, *Ultra-High-Speed Ground Transportation Business Case Analysis*, 6.

22 Washington State Department of Transportation, "Annual Ridership and Historical References, 1993–2014," accessed October 28, 2019, www.wsdot.wa.gov/sites/default/files/2019/03/07/AmtrakCascadesAnnualPerformanceReport2014.pdf.
23 Washington State Department of Transportation, "10-Year Ridership with Historical References, 2009–2018," accessed October 28, 2019, www.wsdot.wa.gov/sites/default/files/2019/03/07/AmtrakCascadesAnnualPerformanceReport2018.pdf.
24 WSP, *Ultra-High-Speed*.
25 WSP, *Ultra-High-Speed*, vii.
26 WSP, *Ultra-High-Speed*.
27 WSP, *Ultra-High-Speed*.
28 Washington State Department of Transportation, "Ultra-High-Speed Ground Transportation Study," accessed October 28, 2019, www.wsdot.wa.gov/planning/studies/ultra-high-speed-travel/ground-transportation-study.
29 US Energy Information Administration (EIA), "Washington State Profile and Analysis," accessed September 30, 2019, www.eia.gov/state/analysis.php?sid=WA; Canada Energy Regulator, "Provincial and Territorial Energy Profiles: British Columbia," accessed October 4, 2019, www.cer-rec.gc.ca/nrg/ntgrtd/mrkt/nrgsstmprfls/bc-eng.html.
30 US Energy Information Administration, "Oregon State Profile and Analysis," accessed October 4, 2019, www.eia.gov/state/analysis.php?sid=OR.
31 EIA, "Washington State Profile."
32 EIA, "Oregon State Profile."
33 Site C Clean Energy Project, "About Site C: Project Benefits," accessed December 2, 2019, www.sitecproject.com/why-site-c/project-benefits.
34 US Energy Information Administration, "Oregon State Profile and Energy Estimates," accessed November 27, 2020, www.eia.gov/state/?sid=OR#tabs-4.
35 On October 15, 2020, Portland General Electric announced that it would close the state's only coal-fired power plant twenty years ahead of schedule. The closure came about because of a 2010 agreement between the utility, regulators, and consumers to reduce air pollution and draw down carbon dioxide emissions. See Kale Williams, "Portland General Electric to Close Only Coal-Fired Power Plant in Oregon," *The Oregonian*, October 15, 2020. Regarding solar, EIA, "Oregon State Profile."
36 EIA, "Washington State Profile."
37 EIA, "Oregon State Profile."
38 Canada Energy Regulator, "Provincial and Territorial Profiles."
39 Canada Energy Regulator, "Provincial and Territorial Profiles."
40 Needham, *Power Lines*, 17.

41 US Department of State, "Columbia River Treaty," accessed October 7, 2019, www.state.gov/columbia-river-treaty/.

42 Northwest Power and Conservation Council, "Northwest Power Pool," accessed October 11, 2019, www.nwcouncil.org/reports/columbia-river-history/NorthwestPowerPool.

43 Northwest Power and Conservation Council.

44 Lawrence A. Dean and Merrill S. Schultz, *Pacific Northwest Coordination Agreement, Background and Issues* (Portland, OR: Northwest Power Planning Council, 1989); and Northwest Power and Conservation Council, "Northwest Power Pool."

45 Washington State Department of Agriculture, "Washington Agriculture," accessed November 1, 2019, https://agr.wa.gov/washington-agriculture.

46 Oregon Department of Agriculture, "Oregon Agriculture: Facts and Figures," accessed November 1, 2019, www.oregon.gov/ODA/shared/Documents/Publications/Administration/ORAgFactsFigures.pdf.

47 British Columbia Ministry of Agriculture, "Agriculture in Brief," accessed November 1, 2019, www2.gov.bc.ca/assets/gov/farming-natural-resources-and-industry/agriculture-and-seafood/statistics/census/census-2016/aginbrief_2016_british_columbia.pdf.

48 US Department of Agriculture, National Agriculture Statistics Service, "Press Release," accessed November 8, 2019, www.nass.usda.gov/Statistics_by_State/Washington/Publications/Current_News_Release/2018/WA_vop.pdf.

49 Washington State Department of Agriculture, "Agriculture: A Cornerstone of Washington's Economy," accessed November 8, 2019, https://agr.wa.gov/washington-agriculture.

50 Oregon Department of Agriculture, "Oregon Agriculture Statistics: September 2020," accessed November 27, 2020, www.oregon.gov/ODA/shared/Documents/Publications/Administration/ORAgFactsFigures.pdf.

51 British Columbia, "2017 British Columbia Agrifood and Seafood International Export Highlights," accessed December 2, 2019, www2.gov.bc.ca/assets/gov/farming-natural-resources-and industry/agriculture-and-seafood/statistics/market-analysis-and-trade-statistics/2017_bc_agrifood_and_seafood_export_highlights.pdf.

52 City of Seattle, *Background Paper, Factors Related to Annexation Policy for Seattle* (Seattle: Department for Community Development,1970).

53 Roger Keil, *Suburban Planet: Making the World Urban from the Outside In* (Cambridge: Polity Press, 2018), 15.

54 Henri Lefebvre, *The Urban Revolution* (Minneapolis: University of Minnesota Press, 2003).

CHAPTER 3

INFRASTRUCTURAL WILDERNESS

Seattle and the Binding of City and Region

THAÏSA WAY AND KEN P. YOCOM

IN HIS BOOK, *THE INVENTION OF RIVERS*, ARCHITECT AND philosopher Dilip da Cunha describes river landscapes through degrees of wetness, terrain defined by water and land.[1] He suggests how the simple mapping of rivers as coherent lines in the landscape with clear edges defies their full extent, what he labels as "rain terrain," an area related to the river that cannot be fully predicted, calibrated, or contained. Rather than depicting the river as a line, da Cunha describes a ubiquitous wetness in which water defines the presence or absence of land. While on the surface he argues for a measuring of terrain through the intensity and relative density of its inundation, his interpretation reveals a more complex inquiry conceiving of land not as whole and static but rather as constantly shifting in its consistency and structure.

Living with the perpetual moistness of the Cascade region, we inherently understand this complexity and recognize its importance in creating and sustaining the biological and environmental richness of the region. In the Pacific Northwest and more specifically the Cascade region, "water is an abiding, if not always appreciated, presence; fog and rain, glacier and waterfall, irrigation canal, and tidal estuary, the Columbia River and Pacific Ocean."[2] While water is a consistent presence defining the characteristics of the region, it has also played a role in shaping the transformative processes of urbanization and city building in the Pacific Northwest. Water is in fact what binds the city and the region, acting as a primary natural system and resource in a hybrid environment that we conceptually identify as infrastructural wilderness. This understanding of place extends the recognition of wilderness as a social and historical construction, advancing the approach of environmental historians to challenge the assumptions for how urban infrastructural systems and their material reaches have been imagined as neutral inscriptions into a preexisting wilderness.[3]

Building further from da Cunha's notion of a river as a rain terrain, the urban landscape can be understood as dynamic, revealing "dense networks of interwoven socio-spatial processes that are simultaneously human, material, natural, discursive, cultural, and organic."[4] Cities are and have always been complex places with long intersecting histories of culture, land, and built fabric, tied together by networks of infrastructure extending from the core to the outermost reaches of a region and beyond. These relationships, shaped through time, comprise convoluted entanglements mediated by emerging technologies of city building and infrastructure. This is nowhere more evident than in Cascadia, where residents claim a regionally distinctive and environmentally privileged approach to urbanization with the emergence of high-tech, sustainably green cities embedded in a wilderness landscape of mountains, forests, and water that is viewed and read as equally engineered and natural.

The Pacific Northwest only became an identifiable region by Euro-American colonizers in the late nineteenth and early twentieth centuries, grounded in perceptions of its remote character coupled with its intrinsic beauty.[5] Seattle as a city at the center of the region was imagined and built with the intention to steward a distinct relationship to nature, one that resists perceived dichotomies of eastern cities. Surrounded by mountains and forests and embraced by water, the metropolis was revered for its setting of magnificent natural beauty while equally disparaged for its pervasive gray and moist skies.[6] As early as the advent of the railroad, circa 1890, Seattle's leaders described the landscape "as one of the features elevating it above eastern counterparts."[7] As described by travel writer Jonathan Raban, Seattle was "the first big city to which people had fled in order to be closer to nature."[8]

Not only did the city of Seattle emerge from the remarkable landscape of the Pacific Northwest, but as geographer Andrew Karvonen argues, it has "long been considered a city in harmony with nature, a metropolis inseparable from and infused with the dramatic and picturesque Pacific Northwest landscape."[9] It is equally known as a technologically advanced "green" city, suggesting that technology and its capacities direct a progressive force for good.[10] Such claims expose an ambivalence of urban residents to fully comprehend the impacts of city building and to recognize that while many came to Seattle for its remoteness and natural beauty, they also exploited the landscape for economic gain. It is an identity deeply tied to its environment, both for the resources therein and as an accessible yet still sublime wilderness. It is this imagined wilderness that is the focus of this inquiry, coveted both for its ability to be harvested for resources and be sustained as untrammeled. It is this landscape that we are calling an *infrastructural wilderness*, a critical element of the urban terrain, that fluid landscape extending from city to territory, from Pike Place Market to the North Cascades National Park and the Skagit River hydroelectric projects. The infrastructural wilderness is thus a landscape that is simultaneously contained within and external to the city, a landscape of imagined nature.

Building on environmental historian William Cronon's foundational thesis of the essential connection between city and hinterland both materially and economically, we suggest a more dependent relationship between city and non-city, between city building and the hinterlands.[11] Building upon three decades of urban environmental histories, we reimagine the idea of wilderness through an urban lens.[12] Specifically, we consider the history of three dams comprising the Skagit River hydroelectric project and how they were inserted into a distant landscape to produce energy needed for the city while both opening access to and retaining culturally embedded perceptions of wilderness. This example reveals a thick web of social, ecological, and technological connections that define Cascadia not as a region of rural, suburban, and urban landscapes but rather as an urban system defying easy distinctions. The insertions of such system-altering technologies into environments is understood as equally "engineered" and "natural," or infrastructural and wilderness. The reading of the perceived place as infrastructural wilderness draws from the ecotechnological system of the city, as described by Matthew Klingle and Thomas Hughes, to consider how it reaches into the hinterlands and landscapes perceived as wilderness.[13] Similar to the concept of a rain terrain, it is composed of two seemingly opposing elements, the urban and the natural. This infrastructural presence percolates from city to wilderness in a dynamic loop shaping both the immediate and distant landscapes and its uses both real and imagined in ways that we rarely grasp, much less tackle.

INFRASTRUCTURAL WILDERNESS

The city is no longer considered an entirely human manifestation distinct from its surroundings, nor confined to politically determined demarcations. Instead, through the processes of urbanization, the city extends across the landscape from parking lots and skyscrapers to the mountains, rivers, and wilderness. Urban historians and political ecologists have increasingly questioned the infrastructural, material, and metabolic processes of cities and urbanization, highlighting the indeterminacies and complexities of metropolitan regions.[14] Such approaches suggest the potential for developing a deeper appreciation of how individual, societal, and cultural practices shape urban landscapes and, specifically, how what we understand as urban infrastructure makes possible both the city and the wilderness as we know them.

Equally limited are the conceptions of wilderness, often conceived as the antithesis to the city. Wilderness has a long history in Western culture and contributes as a foundational idea in the making of the American landscape.[15] Wilderness was once considered sacred as implied in the suggestion of the sublime, described by Edmund Burke.[16] In particular, it is the notion of the sublime that permeates readings of nature in the "wild."[17] With increasing industrialization, by the late 1800s, wilderness and nature were sought as antidotes to the ills of urban progress.[18] Wilderness as an American notion became privileged for the "belief that such wild grandeur would surely give rise to grand literary and cultural achievements."[19]

In Cascadia, and most explicitly in the city of Seattle, one finds the confluence of a technologically driven urbanization and both an ecologically and aesthetically driven preservation of wilderness. The seemingly subtle insertion of hydroelectric infrastructure into the North Cascades range and along the Skagit River reflects this duality of purpose: on the one hand, the dams produce over 33 percent of electricity for Seattle; on the other, the environment created and maintained by the dam is described as a wilderness landscape. This is not untouched wilderness or pristine nature, but, as Linda Nash notes, it is an idea of wilderness that is more in keeping with Richard White's understanding of rivers as "organic machines."[20] It is here that a complex connection between city and surroundings becomes tangible. The infrastructural wilderness is an engineered landscape enabling the city to exist while also retaining our perceptions of wilderness as distinct and external to the city.

THE FLUID TERRAIN OF CITY BUILDING

Early in its development, Seattle was characterized as something quite different from other cities, even those in the Western frontier, often defining itself by the distinct natural resources and landscape. As early as 1871, the city was described by residents as compared to San Francisco: "The climate is better, the water healthier, Sound deeper, trees taller, rivers longer . . . , clams fatter, fish sweeter, [and] mountains higher, . . . a thousand cogent reasons why Seattle should be the garden spot of the earth."[21] Yet from its founding in 1851, its location as a "city in nature" required that "city and nature [be] reconciled" through technocratic and Promethean feats of engineering.[22] Coll Thrush describes Seattle's location as "a bad place to build a city" as settlers and city makers would be required to engage in a form of taming nature by leveling hills and reorienting the flow of water through the landscape in order to build the city they envisioned.[23] To accomplish this, there emerged a necessary municipal governance for such transformative activities.[24] The city's emergence would demand the complex technologies of infrastructure—water, sewers, and electricity—that would change the landscape to improve quality of life while enabling economic development. As Klingle writes, "City engineers employed awesome new technologies—electricity, hydraulic cannons, gasoline-powered dredges—to uplift the environment and its citizens by remaking the city's physical fabric in the name of efficiency, sanitation, and social order."[25]

However, what is often obscured in these descriptions is the critical contribution of the ideas of wilderness as nature. Seattle was built by those seeking access to available natural resources, bent on extracting from the hinterlands, first the trees and fish and then gold. And yet it was equally appreciated for its awe-inspiring setting. While such descriptions privilege land, water is just as powerful and present. Framed within this context, Cascadia's rivers and corresponding terrain are essential in shaping a regionally coherent conception of the complex relationships between the ideas of city, nature, and wilderness, both perceived and real. Rivers host an intrinsic beauty while offering the power to transform the landscape. These rivers are at the core of many ecological narratives in the region that recall the sight of anadromous salmon so dense that you could "walk across the backs from one bank to the other."[26] While difficult to confirm, the message clearly speaks to the historically rich productivity of the region's landscape.

Nevertheless, through periods of colonization and subsequent development, many of the rivers were transformed through intense socio-technological processes. Like the Duwamish, they were constrained, polluted, and redirected. Some, such as the Black River that once flowed out of Lake Washington, were lost altogether. The rivers were reconstructed as White's "organic machines," systems neither entirely mechanical nor organic but bound and engineered both by natural forces and political economies of power and labor.

In order to harness the power of these organic machines, public entities and individuals alike constructed dams. More than 1,150 dams have been constructed in Washington since the mid-1850s, primarily for flood reduction, irrigation, and energy production.[27] Such dams serve as socio-technological entities deeply embedded in larger political processes yet are understood as static structures impounding the forces and restraining the rhythms of a river's flow.[28] However, dams are equally dynamic, dependent on regional precipitation rhythms in their operational capacities for focusing, harnessing, and converting the energy contained in their aqueous flows.

For some, the dams have dampened the majesty and wildness of frothing cascades and deep aquamarine pools; for others, they have harnessed the power of rivers, capitalizing on the energy derived from consistent flows to reliably support the urbanization of the region, providing the lowest energy costs of anywhere in the country. One series of three such dams along the Skagit River offers the opportunity to consider how these seemingly conflicting narratives have in fact been woven together to create an example of what we define as infrastructural wilderness.

The Skagit River in northern Washington is a critically important riverine landscape. It has been engineered for energy production, in turn opening access to a once remote wilderness landscape. With three dams in its upper reaches built by 1949, the dams rise in the midst of the North Cascades national forests and parks, with stunning glacier-capped mountain peaks reflected in the aquamarine waters of their reservoirs. The area attracts hikers, bikers, skiers, kayakers, boaters, fishers, and backcountry adventurers to its notable rock cliffs, glacier slopes, and trail networks. The dams are integral to this landscape, creating the large reservoirs, including 37-km-long Ross Lake, and access through road infrastructure to this perceived wilderness a short few hours' drive from downtown Seattle.

A LAND OF CITY AND WILDERNESS

One of the largest drainages to the Salish Sea, the Skagit River originates in the southern reaches of the Canadian Cascades, flowing predominantly south and west through a rugged landscape of steep gorges and glacial valleys before cutting a path through the arable lowlands of the Skagit Valley and emptying in the shallows of Skagit Bay on the eastern shores of the Salish Sea. The name derives from the two tribes of the Lushootseed people, upper river and lower river, who inhabited the watershed since shortly after the retreat of the Fraser glaciation. These tribes cultivated the landscape in ways historians are only beginning to recognize.

The Washington Territory opened to homesteading in 1846, although the Skagit River area saw European settlers starting only in the 1870s and later. With these settlers came new uses for the land.[29] Settler migrations up the Stehekin, Cascade, and Skagit Rivers continued through the 1880s with homesteaders setting up as traders and shopkeepers, as well as innkeepers, serving the trappers, miners, and prospectors who worked the rivers with high hopes albeit little success. Due to the rugged character of the large rivers, tall mountains, and deep forests, the land nevertheless remained more wild than overtly cultivated or settled.[30] The first wagon road was built into the area in 1892, although it was not until 1972 that the North Cascades Highway, Route 20, opened access over the mountains from May to early November.

Early in its European settlement, the area was recognized as a place of significant potential for extraction (mining, logging, fishing, etc.) yet one that might be preserved as wilderness.[31] A petition to create a national park was submitted in 1892, just seven years before Mount Rainier was so designated (the fifth in the nation). In 1897, a Forest Reserve was designated, although it would take until 1968 before the North Cascades was named a national park alongside the Ross Lake and Lake Chelan National Recreation Areas. The US Congress established North Cascades National Park Service Complex to preserve "certain majestic mountain scenery, snowfields, glaciers, alpine meadows, lakes and other unique glaciated features." In 1988, the Washington Park Wilderness Act designated 634,614 acres as the Stephen Mather Wilderness Area. Today more than 240 km in length and draining an area of nearly 7,000 km^2, much of the landscape of the river, its tributaries, and their contributing land is publicly owned and managed, officially demarcated as National and Provincial Park, National Recreation Area, and Wild and Scenic as well as a critically important hydroelectric project.

While tempting to imagine this region as untrammeled wilderness until the opening of Highway 20, still the only road that crosses west to east in the North Cascades, this was not the case. Between 1919 and 1949, the Skagit River, one of the region's three major rivers, was dammed in order to provide hydroelectrical power for the city of Seattle. As Nash writes, "These material changes unquestionably shaped human experience of the river, but the material changes themselves proceeded in part from changes in language and forms of mediation."[32] The dams were, on the one hand, productive as they enabled the flow of power, making Seattle's growth possible, but they also dramatically changed the river's flows and destroyed salmon-spawning grounds. Yet, as the river was being significantly engineered, there was an equally audacious effort to retain its "natural beauty" as a distinct resource for urban dwellers, suggesting that the river could be both an infrastructure for energy flow and a revered landscape of wilderness that enticed so many to settle and colonize the Pacific Northwest.[33]

So how did the Skagit River's three dams come to be built? Upon the successful construction and operation of the nation's first municipally owned hydroelectric project on the Cedar River in 1905 to supply power for city-owned streetlights, demand for electricity increased to a point that well surpassed the capacity of the single dam. A race to secure the rights for generation of electricity ensued, playing out in public meetings, at voting halls, and behind closed doors, pitting public and private entities in competition.[34] Meanwhile, Seattle established a new municipal agency, Seattle City Light, appointing engineer J. D. Ross (1872–1939) as superintendent. A proponent of municipal control of power generation and the electrical network, Ross, in Promethean fashion, convinced the city council and lobbied the federal government to provide the funds for construction of three dams on the Skagit River—Gorge High, Diablo, and Ruby (later renamed Ross) dams—more than 160 km to the northeast of the city in the heart of the Cascade mountain range (see figure 3.1).[35] Building the requisite infrastructure to access this remote location and the construction of the dams, powerhouses, and distribution networks started in 1919 and continued for the following three decades.[36]

For Ross, the construction and subsequent creation of electrical power for the city from the remote hydroelectric project was part of a larger city building enterprise that would establish Seattle as the preeminent city on the West Coast. As he wrote, "Cheap power is the key to the future of industrial supremacy and will make the Puget Sound Basin one of the World's busiest workshops."[37] William Sparks argues that Ross "was not a mere visionary. He successfully combined a basic disposition toward humanitarianism with technical ability. The result was a philosophy of social engineering . . . a panacea which would right existing wrongs and cure dozens of social ills."[38] To advocate for the project and to promote the strong connection between city and hinterland, Ross inaugurated tours up the Skagit River valley, urging that the dams need not serve solely as a source of energy but also as a destination of human ingenuity and technological prowess set within the remote and dramatic landscape of the North Cascades. The Skagit River Railway was constructed to simultaneously support supply lines for the projects and open public access to the relatively uninhabited mountains and old-growth forest. Eventually the lines included a cable-powered tramway lifting visitors 313 feet up the side of the canyon, across the reservoir created by the Diablo Dam, and finally to the top of Ross Dam by train.

Throughout the development of the hydroelectric projects, Ross sought to entice visitors and showcase the land. Initially he transformed small areas immediately adjacent to the dams into garden areas, stating, "The beautification of the environments of the Skagit Power Plant, added to the majestic setting of its mountains and river, makes it one of the principal beauty spots of the World. Plants,

FIGURE 3.1. Diablo Dam, the second of three dams built as part of the Skagit River hydroelectric project, owned and operated by the City of Seattle. The dams produce approximately one-third of the electricity used in Seattle today. Courtesy of City of Seattle.

animals and birds are being brought in from all parts of the world to add to the scenic grandeur of the valley."[39] His ideas suggested a landscape that would provide evidence of the bounty of the region and, similar to contemporary botanical gardens and zoos, could host flora and fauna from around the globe. Ross introduced exotic species ranging from black squirrels to albino deer while reintroducing the American beaver that had been locally extirpated and importing mountain goats to bolster a waning population. Furthermore, he began programs to stock the reservoirs behind the dams with trout in efforts to create a viable fishery to attract anglers. Though unrealized, he had plans to transform a hillside by planting ten thousand rhododendrons and dogwood trees to provide an abundance of flowers in contrast to the dense green of the surrounding coniferous forests as backdrop for the visitors viewing the original powerhouse. Summer tours of the dams were initiated in 1926, and by 1937, 25,000 people visited each year.[40]

However, Ross's vision of a more exotic landscape proved to be an expensive endeavor and eventually succumbed to a more regionally defined identity, one that would resonate with the perception of the landscape as a native wilderness, a view that was core to the region's emerging identity. The view of mountains, streams, and native forests were designed to offer inspiring vistas of the engineered dams framed by a correspondingly awe-some experience of a glacial panorama. Today, the region annually receives nearly a million visitors who come to recreate in the backcountry and wilderness.[41]

The distinct design of the Skagit River dams is evident when compared with contemporary dam landscapes. On the other side of the country, the Tennessee Valley Authority (TVA), established in 1933, was an apt model for energy production landscapes in public ownership that defined locations as cultivated places of production. Resulting in more than twenty-nine hydroelectric dams in the Tennessee Valley, the dams were intended not only for flood protection and energy production but also to build a regional economy in an area hard hit by the Great Depression. The project's architect, Roland Wank, took the engineer's design for the Norris Dam and treated "the dam and its powerhouse as a new kind of modern sculpture."[42] Beautiful parkways led to the dams, lawns with modernist tourist centers ideal for viewing the dam site. Furthermore, these dams produced energy to support a rural population, one that was intended to demonstrate the best of the American farmer and laborer of the land. Accordingly, the dam landscapes were presented as set in equally stewarded and managed landscapes, even extending into the apparent forests, planted as managed plantations with regional species introduced to maximize the aesthetic character of the landscape.[43]

The Skagit River dams are not delineated by lawns or manicured gardens as Ross imagined. The mountains drop to the reservoir edges, native forests surround the dams as if they had been merely slipped into an otherwise pristine wilderness (see figure 3.2). Rather than a realization of Ross's dream of a manicured Eden, the dams lie amid a vast wilderness, as valued by Seattle's and Cascadia's leaders and public. The relationship between hydroelectric use and "some of the country's wildest, most rugged, and beautiful mountains" represents a contested terrain, a political geography, pitting urban needs against wilderness preservation.[44] Yet what has emerged in the twenty-first century is a seemingly seamless integration of the two purposes, the result of almost a century of development of both the landscape and the narrative.

Dams have empowered a "simultaneous process of social and environmental change in which new kinds of space are created and destroyed, ranging from the technological networks that give sustenance to the modern city to new appropriations of nature within the urban landscape."[45] While the dams composing the Skagit hydroelectric project have transformed the wilderness area, that is not the message that is forefronted at the actual sites. Rather than the TVA's visitor centers, there is the North Cascades Environmental Learning Center where visitors learn about nature, staying in cabins tucked into the forest. Museum collections include 700,000 scientific specimens, archaeological and historical artifacts that evidence the significance of the region's natural and cultural histories. In addition, visitors can take a boat tour to experience "wildlife habitat and enjoy views of snow-clad mountain peaks, islands, and waterfalls," as opposed to a round on the TVA's golf courses.[46]

While the nearby town of Diablo was built to house those constructing and managing the dams, it was not a model town like Norris, Tennessee. It was a simple outpost with enough amenities to serve the community, including access to hiking trails that would take one to hard-to-reach fire posts and viewpoints. It was on such fire posts that legendary artists and writers such as Jack Kerouac and

FIGURE 3.2. Ross Dam and reservoir set within the designated wilderness of the North Cascades National Park Complex. Courtesy of City of Seattle.

Gary Snyder, while serving as fire watchers, documented the landscape through poetry and prose.

This dual role of the dams as infrastructure for Seattle and the gateway to backcountry wilderness challenges the assumptions that city and hinterland act as distinct systems merely engaged in transactional relationships. Instead, the relationship of the Skagit River, its engineered dams alongside its remarkable national park, to the metropolis of Seattle offers evidence of how urban infrastructure is deeply embedded in the wilderness, describing a distinct place, an infrastructural wilderness. It is a place that is simultaneously urban and non-urban. The awe of the visit is both the sublime experience of viewing and possibly exploring the gorges, mountains, and river *and* that it is a major power generator for the city of Seattle, over 160 km away. This is a site of production celebrating its role in supporting the infrastructure of the city and wilderness alike.

THE BINDING OF CITY AND REGION

Fundamentally, water and its associated rain terrain are critical to the development and capacity of urban regions. The Skagit River hydroelectric project and conveyance infrastructure are examples of the complex socio-ecological entanglements of material and social domains that bind the city and region. The Skagit River dams expanded the city's control of distant natural processes and hinterland by appropriating their flows for the creation of electrical power and remain vitally important for regional conceptions of Seattle as a growing metropolis of economic importance in the far corner of the nation. However, the dams and resulting distribution network obscured the complexity and immense scale of the natural and socio-technological processes required to develop and maintain the power system that progressively bound the city in its resource dependency to the wilderness areas via a tangible infrastructural network of dams, turbines, power converters, and wires.

This relationship is not unique to Seattle. Over 84,000 dams have been built in the United States, transforming the landscape for the technocratic needs of nation building and city making. Lewis Mumford described the technological advancements of dams as blurring the perceived boundaries between society and nature, suggesting a socio-technological world of organic machines and ecological processes that realize the potential of human ingenuity and natural productivity.[47] Today we understand how dams have radically reshaped territories, from the erasures of tribal lands to site reservoirs, to the cultivation of former dry lands with irrigated agriculture, to the electrification of cities. Dams have been critical to the political economy of power that has spatially reconstrued urban systems and the relationships of the seemingly urban to the landscape of the hinterlands.

On the one hand are the ways that emerging understandings of nature in the urban environment are manifesting. On the other, appropriations of urban systems in the so-called natural landscape have been largely ignored, caught between the banality of infrastructure and the majesty of wilderness. Kees Lokman frames dams within the context of the Anthropocene, as systems "where human and natural processes have become . . . simultaneously highly engineered and indeterminate in its use(s), creating a landscape of dynamic stability, continually manufacturing new ecologies and economies."[48] The fluidity of water and energy of the dam, its river, and its landscape link city and region in ways that can essentially only be read as one coherent and yet dynamic landscape serving multiple and diverse needs and purposes.

Today, the hydroelectric projects on the Skagit River remain geographically distant from the city and are essentially inaccessible in winter months, yet power is always moving directly from the wilderness to the city. The distance challenges how the concept of a sublime wilderness distinct from urban influences is transformed to recognize the thick interdependencies of the two landscapes: urban and wilderness. The nebulous boundaries between city and wilderness are transcended through an infrastructure of electrical generation, distribution, and access. In Seattle, and more broadly within Cascadia, there emerges a distinct relationship between the flows of urbanization informed by the equally powerful forces and flows of wilderness. Seattle's development suggests how the far reaches of the region have and continue to be informed by the city while revealing significant agency of the wilderness landscape in the perception and experience of the urban nature of Cascadia. The legacy of the dam infrastructure on the health and abundance of fish and other species native to the river landscape has yet to be fully understood, an investigation that is questioned by local tribal members and will be important to the future sustainability of urban systems. The North Cascades and the Skagit River are not merely hinterland to the metropolis but rather equal and essential partners in the production and perception of the urban landscape. This is not a neutral condition but rather, as with any cyborg, a technological creation that must now be managed and cared for in order to produce a future that expands and enhances the health and well-being of the city and its hinterland.

NOTES

We acknowledge the land and waters we describe in this chapter as the traditional home of the Original Peoples of the Coast Salish region. Without them, we would not have access to this knowledge and dialogue. We thank the original caretakers of this land, who are still here.

1 Dilip da Cunha, *The Invention of Rivers: Alexander's Eye and Ganga's Descent* (Philadelphia: University of Pennsylvania Press, 2019).

2 G. Thomas Edwards and Carlos A. Schwantes, eds., *Experiences in a Promised Land: Essays in Pacific Northwest History* (Seattle: University of Washington Press, 1986), 3.

3 See Paul S. Sutter, "The World with Us: The State of American Environmental History," *Journal of American History* 100, no. 1 (2013): 94–119.

4 Erik Swyngedouw and Maria Kaika, "The Environment of the City . . . or the Urbanization of Nature," in *A Companion to the City*, ed. Gary Bridge and Sophie Watson (Malden, MA: Blackwell, 2000), 568.

5 John M. Findlay, "A Fishy Proposition: Regional Identity in the Pacific Northwest," in *Many Wests: Place, Culture, and Regional Identity*, ed. David M. Wrobel and Michael C. Steiner (Lawrence: University Press of Kansas, 1997), 37–70.

6 See Roberto Brambilla, *Learning from Seattle: What Makes Cities Livable?* (New York: Institute for Environmental Action, Partners for Livable Places, 1979); Findlay, "Fishy Proposition"; Andrew Karvonen, "Metronatural™: Inventing and Reworking Urban Nature in Seattle," *Progress in Planning* 74, no. 4 (2010): 153–202; and Jonathan Raban, "The Next Last Frontier: A Newcomer's Journey through the Pacific Northwest," *Harper's* 287, no. 1719 (1993): 30.

7 Findlay, "Fishy Proposition," 53–54.

8 Raban, "Next Last Frontier," 33.

9 Karvonen, "Metronatural™," 153.

10 Matthew Klingle, *Emerald City: An Environmental History of Seattle* (New Haven, CT: Yale University Press, 2007); and Andrew Karvonen and Ken Yocom, "The Civics of Urban Nature: Enacting Hybrid Landscapes," *Environment and Planning A* 43, no. 6 (2011): 1305–22.

11 William Cronon, *Nature's Metropolis: Chicago and the Great West* (New York: W. W. Norton, 1991).
12 Thaïsa Way, "What Is the Urban Landscape and What Role in Urban History?," *Journal of Urban History* 45, no. 3 (2019): 595–600.
13 Sutter, "World with Us," 115. See also Klingle, *Emerald City*, 5; and Thomas P. Hughes, *Human-Built World: How to Think about Technology and Culture* (Chicago: University of Chicago Press, 2004), esp. 153–74.
14 See Maria Kaika, *City of Flows: Modernity, Nature, and the City* (New York: Routledge, 2004); Nik Heynen, Maria Kaika, and Erik Swyngedouw, *In the Nature of Cities: Urban Political Ecology and the Politics of Urban Metabolism* (London: Routledge, 2006); Anne Rademacher, "Urban Political Ecology," *Annual Review of Anthropology* 44, no. 1 (2015): 137–52; Henrik Ernstson and Erik Swyngedouw, eds., *Urban Political Ecology in the Anthropo-Obscene: Interruptions and Possibilities*, Questioning Cities Series (London: Routledge, 2019); and Matthew Gandy, *The Fabric of Space: Water, Modernity, and the Urban Imagination* (Cambridge, MA: MIT Press, 2014).
15 Roderick Frazier Nash, *Wilderness and the American Mind* (New Haven, CT: Yale University Press, 1967).
16 Edmund Burke, *A Philosophical Enquiry into the Origin of Our Ideas of the Sublime and Beautiful: With an Introductory Discourse Concerning Taste, and Several Other Additions* (1757; Montrose, UK: Printed by D. Buchanan, 1803); and William Cronon, ed., *Uncommon Ground: Toward Reinventing Nature* (New York: W. W. Norton, 1995).
17 Cronon, *Uncommon Ground*; and William Cronon, "The Trouble with Wilderness; Or, Getting Back to the Wrong Nature," *Environmental History* 1, no. 1 (1996): 7–28.
18 Nash, *Wilderness*.
19 Bryan McDonald, "Considering the Nature of Wilderness: Reflections on Roderick Nash's Wilderness and the American Mind," *Organization & Environment* 14, no. 2 (2001): 192.
20 Linda Nash, "The Changing Experience of Nature: Historical Encounters with a Northwest River," *Journal of American History* 86, no. 4 (March 2000): 1600–1629; and Richard White, *The Organic Machine: The Remaking of the Columbia River* (New York: Hill and Wang, 1995).
21 Edwards and Schwantes, *Experiences in a Promised Land*, xiv.
22 Karvonen, "Metronatural™." For the Promethean worldview as a rationalized perspective, see John S. Dryzek, *The Politics of the Earth: Environmental Discourses* (New York: Oxford University Press, 1997); Kaika, *City of Flows*; Maria Kaika, "Dams as Symbols of Modernization: The Urbanization of Nature between Geographical Imagination and Materiality," *Annals of the Association of American Geographers* 96, no. 2 (2006): 276–301; and Karvonen, "Metronatural™," 160–61.

23 Coll Thrush, "City of the Changers: Indigenous People and the Transformation of Seattle's Watersheds," *Pacific Historical Review* 75, no. 1 (2006): 89.
24 Klingle, *Emerald City*; and Karvonen, "Metronatural™," 156–59.
25 Klingle, *Emerald City*, 5.
26 On salmon, see David R. Montgomery, *King of Fish: The Thousand-Year Run of Salmon* (Boulder, CO: Westview Press, 2003); Joseph E. Taylor, *Making Salmon: An Environmental History of the Northwest Fisheries Crisis* (Seattle: University of Washington Press, 1999); and Jim Lichatowich, *Salmon without Rivers: A History of the Pacific Salmon Crisis* (Washington, DC: Island Press, 1999).
27 US Army Corps of Engineers (USACE), "National Inventory of Dams," accessed November 6, 2019, http://nid.usace.army.mil.
28 Christine Macy, *Dams*, Norton/Library of Congress Visual Sourcebooks in Architecture, Design, and Engineering (New York and Washington, DC: W. W. Norton, Library of Congress, 2010); and Kees Lokman, "Dam[ned] Landscapes: Envisioning Fluid Geographies," *Journal of Architectural Education* 70, no. 1 (2016): 6–12.
29 Lawrence Boome, "Personal Perspectives on Indian History in the Skagit River Valley," in *Of Man, Time, and a River: The Skagit River, How Should It Be Used? Proceedings of a Series of Meetings Held in the Skagit River Valley, September–October 1976*, ed. Roland L. De Lorme, Center for Pacific Northwest Studies, Occasional Paper 10 (Bellingham: Western Washington State College, 1977), 6–7.
30 Charles Dwelley, "Early Days on the River," in De Lorme, *Of Man*, 11–15; and Nash, "Changing Experience."
31 Paul C. Pitzer, *Building the Skagit: A Century of Upper Skagit Valley History, 1870–1970* (Portland: Galley Press, 1978).
32 Nash, "Changing Experience," 1603–4.
33 William O'Dell Sparks, *J. D. Ross and Seattle City Light, 1917–1932* (master's thesis, University of Washington, 1964), 183.
34 Myra L. Phelps, *Public Works in Seattle: A Narrative History [of] the Engineering Department, 1875–1975* (Seattle: Engineering Department, 1978), 138–47.
35 Sparks, *J. D. Ross*, 75–109; and Klingle, *Emerald City*, 92.
36 Pitzer, *Building the Skagit*, 23–59.
37 Ross 1927, letter, as cited by Sparks, *J. D. Ross*, 4.
38 Ross 1927, letter, as cited by Sparks, *J. D. Ross*, 23.
39 City of Seattle 1935, as cited by Nash, "Changing Experience," 53.
40 City of Seattle 1937, as cited by Pitzer, *Building the Skagit*, 67.
41 Vince Richardson, "Nearly 1 Million Visitors Stopped by North Cascades Park Complex in 2016," *Seattle Times*, April 6, 2017.
42 Tennessee Valley Authority, "Design for the Public Good," accessed March 24, 2020, www.tva.gov/About-TVA/Our-History/Built-for-the-People/Design-for-the-Public-Good.

43 Avigail Sachs, personal communication to Thaïsa Way, December 6, 2019.

44 David Louter, *Contested Terrain: North Cascades National Park Service Complex: An Administrative History* (Seattle: National Park Service, 1998), 1.

45 Gandy, *Fabric of Space*, 63.

46 City of Seattle, "Diablo Lake Boat Tours," accessed December 04, 2019, www.seattle.gov/light/damtours/diablotours.asp.

47 Lewis Mumford, "The Architecture of Power," *New Yorker*, June 7, 1941, 58.

48 Lokman, "Dam[ned] Landscapes," 10.

CHAPTER 4

URBANIZATION AND WATER GOVERNANCE DYNAMICS IN BEND AND HOOD RIVER, OREGON

ALIDA CANTOR AND ALEXANDER REID ROSS

THE QUESTION OF WHAT IS AND IS NOT "URBANIZED" MAY seem obvious at first consideration: urban areas are large, crowded metropolitan cities, which stand in contrast to their quieter, rural, country counterparts. Yet, as Ash Amin and Nigel Thrift point out in the following passage, these spaces are not so easily delineated:

> The city is everywhere and in everything. If the urbanized world now is a chain of metropolitan areas connected by places/corridors of communication (airports and airways, stations and railways, parking lots and motorways, teleports and information highways), then what is not the urban? Is it the town, the village, the countryside? Maybe, but only to a limited degree. The footprints of the city are all over these places, in the form of city commuters, tourists, teleworking, the media, and the urbanization of lifestyles. The traditional divide between the city and the countryside has been perforated.[1]

Increasingly, spaces such as suburbia and exurbia render such categories less and less meaningful. Commuters may live in the country and commute or telecommute daily to jobs in a city. The city and the country make each other possible through resource flows. City residents rely upon water, food, and energy produced in rural areas, and even in rural areas, amenities such as Starbucks coffee shops are likely not far away. Scholars from human-environment geography and related fields have argued that the seemingly discrete spaces of the country and the city are interconnected, interdependent, and difficult to distinguish into tidy and separate categories.[2] From this perspective, rural and urban landscapes can and should be seen as coproduced—that is, they shape each other in inextricable ways in dynamic processes of urbanization.

In this chapter, we examine urbanization beyond the borders of "Mainstreet Cascadia," focusing on the politics of water management in two smaller Oregon towns located in communities outside the region's urban core: Bend and Hood River (see map I.1 in introduction).[3] The chapter draws primarily from interviews with water managers, environmental advocates, farmers, community members, and other stakeholders involved in water governance in the Bend and Hood River areas.[4] After discussing existing scholarship on urbanization and exurban political ecology, we examine urbanization trends in Bend and Hood River, then discuss how population growth and demographic changes in these communities have impacted water resource allocation and governance.

Contrary to expectations, we find that due to the relatively small proportion of water dedicated to urban residential use compared to agricultural water use, population growth has not increased demand for water in ways that overburden available water resources. However, growing populations have impacted water governance and management in both Bend and Hood River in a number of ways, as new residents bring with them new sets of expectations, experiences, and values. We situate these changes within regional and global contexts to show how local-scale water governance dynamics are tied to broader urbanization processes.

URBANIZATION, URBAN METABOLISM, AND EXURBIA

Urbanization includes the landscapes and processes of resource production and consumption that are involved in the circulation of energy and materials, known as "urban metabolism." The urban metabolism of cities relies upon an extensive regional and global network of resources for food, energy, water, and other materials that are integral to the form of the city itself; meanwhile, rural places are integrated into urban landscapes through the provision of these materials as well as the absorption of waste.[5] Urban metabolism is a useful tool for exploring relationships between country and city. For example, environmental historian William Cronon describes the relationships between the growth of the city of Chicago and the simultaneous development of the rural Midwest, and geographer Richard Walker notes the seemingly rural characteristics, such as open space and greenery, that exist within the city of San Francisco as well as the landscapes of urbanization processes, such as reservoirs and dumps, that develop in the countryside.[6] Less tangible goods are also implicated in these flows of resources. Urban areas are often seen as an engine of culture and economic growth, whereas rural areas provide ecosystem services and bucolic recreation landscapes. People migrate within and between these interconnected spaces, bringing with them their cultural values, ideals, and assumptions. Urban and rural landscapes that are distinctive in character, yet fundamentally interconnected, emerge through these flows of resources, people, and ideas. Approaches involving urbanization and urban metabolism have been particularly useful in considering water resource management, since cities frequently draw upon broader territories for water supply.[7]

The lens of "extended urbanization" also provides a useful way to analyze urbanization beyond city boundaries, including the growth of extended suburban and exurban landscapes.[8] As global economic trends increasingly move manufacturing and resource extraction overseas and the US economy shifts toward service and technological sector jobs, increasingly decentralized and hybrid post-industrial landscapes known as exurbia emerge at the rural-urban interface.[9] Exurbia is characterized by low-density residential development in rural areas, particularly by people who have left denser cities seeking a rural lifestyle but one with modern conveniences and services.[10] Developing exurban areas face challenges such as population growth, demographic shifts, climate change, and new demands on land use. Through these dynamic shifts, rural economies based on resource extraction can come into conflict with new arrangements organized around selling in-migrants the experience of a reconstructed rural place.[11] This process of urbanization into spaces coded "rural" can create friction between groups perceived as newcomers and old-timers—categories that scholars have pointed out can be overly simplistic and are in reality less coherent than they may initially appear.[12] Be this as it may, in some cases, long-time residents and their livelihoods may indeed clash with incoming residents who view exurbia as something like *terra nullius*, or a kind of blank slate onto which they might inscribe their own ways of life.[13] In other cases, these interactions and new arrangements can lead to interesting and sometimes unexpected collaborations.[14] Environmental conflicts in exurban places of the western United States can take shape over numerous aspects of land use changes—including disputes related to the governance and use of water.[15]

URBANIZATION IN BEND AND HOOD RIVER, OREGON

While the larger cities of Mainstreet Cascadia, such as Seattle and Portland, dominate ideas of the urban Pacific Northwest, Bend and Hood River are two smaller cities within the Cascadia region that have grown rapidly over the past several decades (see table 4.1).

Urban growth trends in Oregon accelerated when escalating urban real estate prices and the crippling drought that struck California from 2011 to 2017 drew Californians northward toward lower-cost options such as Oregon.[16] As Portland has become an increasingly popular place to live, real estate prices have increased there as well, pushing current and potential residents to other areas in the state, including smaller cities such as Bend and Hood River. Approximately sixty miles east of Portland, the smaller city of Hood River has

TABLE 4.1. Population growth in selected cities in Oregon, 1990 to 2017

CITY/STATE/ COUNTRY	1990 POPULATION	2017 POPULATION	POPULATION CHANGE
Bend, OR	24,024	94,520	+293%
Hood River, OR	4,677	7,686	+64%
Portland, OR	487,849	647,805	+33%
Oregon State	2.85 million	4.14 million	+45%
USA	250 million	327 million	+31%

Source: Data from US Census.

absorbed some of Portland's population growth. Hood River has experienced subsequent changes to rural lifestyles, priorities, values, and ideas in the erstwhile-rural area. Meanwhile, the medium-sized city of Bend in central Oregon is one of the fastest-growing cities in the region. Bend has become a new hub for those seeking the trappings of a cosmopolitan urban life, including a rapidly expanding craft brewery industry, with the amenities of a rural place, such as hiking, fishing, and scenic panoramas. Bend and Hood River, both popular destinations for retirees, have undergone sharp increases in real estate prices since the end of the 2008 economic recession. While both Hood River and Bend are defined by the US Census as urban, they fit our earlier definition of exurbia in that they are smaller communities located in rural settings that have experienced rapid growth largely driven by amenity migrants seeking both a rural lifestyle and modern urban conveniences.[17]

Despite the growth described above, both the Hood River and Bend areas retain agricultural production and identity today. Hood River County relies on its myriad pear, apple, and cherry orchards for the majority of its economic productivity, while the Deschutes River Basin produces a variety of agricultural products, including nearly half of the world's carrot seeds. Whether processed or shipped in original form, these commodities circulate through the global market, facilitating agricultural production in other locations (for instance, carrot farming) and consumption in still others, allowing for the maintenance of urban livelihoods far removed from the process and point of food production. This tightly interwoven global network of food production and consumption sustains both regional economies throughout Cascadia and global cities reliant on agricultural imports. This production system is made possible through significant quantities of water. Irrigated agriculture represents nearly 80 percent of overall water use in Oregon, which mirrors patterns of agricultural water use more broadly across the western United States.[18]

Despite the ongoing importance of agricultural production, both Hood River and Bend have followed a broader movement away from primary resource production, in particular the timber industry, from the 1980s to the present. Hood River largely moved toward windsurfing and recreation as a substitute for lost timber industry dollars. Bend's timber interests invested in revamping the area's infrastructure to foster the atmosphere of a resort and tourist destination east of the Cascades.[19] Urbanization trends in Bend and Hood River have also gained a foothold through tech industries, including the transformation of Bend into a bedroom community for Silicon Valley as tech workers take advantage of Bend's cheaper housing combined with telecommuting and the short seventy-minute flight between Bend and the Bay Area. Just outside of Bend, the small town of Prineville has recently seen the development of data centers for Facebook and Apple; meanwhile, a corporation making drones has been established in Hood River. In both Bend and Hood River, real estate speculation, the developing technological industry, and recreational amenities and the service industry surrounding them have contributed to the movement from extractive resource markets to "post-industrial" economies. At the same time, timber and agricultural production do still persist in the areas surrounding both of these communities, shaping regional identities and cultures as well as water use.

DECOUPLING URBAN GROWTH AND WATER USE

One might reasonably assume that such rapid population growth is inevitably paired with a rapid increase in water use. New residents require water for drinking and household use; with new housing developments come new lawns and plumbing connections. Moreover, rapidly expanding industries such as microbreweries (Bend boasts one of the highest rates of microbreweries per capita in the United States) require water as well. Yet, contrary to this assumption, the total water use in metropolitan Bend and Hood River has actually remained fairly steady in the face of rapid population growth, representing a decline in per capita water use over time. This trend of "dematerialization" has been described by resource geographers who note that resource use is increasingly decoupled from population and economic growth.[20]

Bend, for example, added over ten thousand new water meters between 2000 and 2017. Yet the actual average daily demand for water in Bend barely ticked up. Area water managers attribute this to conservation efforts, including public-facing water conservation programs as well as efforts such as improvements to water metering and repairing leaky infrastructure. As a water manager explained, "I know our indoor [water] use is going down. Our trend is down.

Breweries are less than 2 percent of our annual supply; they don't use a lot of water."[21]

To understand this dynamic in more depth, it is crucial to contextualize city water use within a broader picture of water use, which in the case of Bend means looking at the entire Deschutes River Basin. Within the Deschutes River Basin, cities actually use far less water overall, and less water per acre on average, than agricultural activities. As an environmental advocate explained:

> A lot of people think that Bend's growing, so the water pressure must be intense . . . but when you look at the municipal use of water, it's 2 percent of the water budget in the Deschutes Basin. So, it's an important but very small component of water use. And 86 percent of the water rights are controlled by irrigated agriculture. So, kind of counterintuitively, urbanization of what used to be agricultural land actually creates a water supply opportunity.[22]

As noted above, this division of water supplies echoes water use patterns across the arid western United States more broadly, where irrigated agriculture accounts for the vast majority of water use.

Meanwhile, in recent years, several species in the Deschutes River Basin, including the Oregon spotted frog, bull trout, and steelhead, have been listed as threatened under the federal Endangered Species Act. These endangered species require significantly higher instream flows than currently exist. The spotted frog, in particular, requires about six times more water than is currently released through the Upper Deschutes River in winter. As a water manager explained:

> [The city of Bend] gets about 14,000 acre-feet in a typical year.[23] We identified . . . about 16,000 acre-feet of water [needed] for municipal needs over the next fifty years. And then, just to get the instream flow targets met [for endangered species] . . . we needed about 130,000 acre-feet. So . . . clearly it couldn't come from the cities. It seems there's a lot of opportunity on the [irrigation] district side for efficiencies to produce that amount of water. That's sort of a simple conclusion there.[24]

That is, even if Bend and the area's other cities stopped using water altogether, they would not save enough water to meet the higher instream flow targets required to meet the needs of the endangered species because agricultural irrigation uses so much more water than cities.

This examination of the relationships between urbanization, agriculture, and water use reveals several important and interconnected points. First, urbanization is not increasing overall water use in the Bend area, because of municipal conservation efforts as well as the relatively small percentage of water used for cities as compared to agriculture in the region. Second, despite the changing economy and rapid urban development, agriculture remains important when it comes to water and is still the largest overall water user. Third, tensions around water use exist, especially in terms of meeting the instream flow needs for endangered species. Because agriculture represents the biggest water user, there is increasing pressure on agricultural irrigation districts to conserve more water to meet instream flow targets.

EXURBAN DEVELOPMENT AND LAND/WATER USE TENSIONS

Hood River is a historically agricultural community that exemplifies many tensions around land and water use associated with exurbia and urbanization. In Hood River, like in Bend, the small city's growing population has not driven increased water demands to the point of stressing the river basin. Instead, tensions in this case have emerged from the territorial encroachment of residential populations and new development onto irrigation districts. Hood River has an urban growth boundary that effectively mitigates residential sprawl into the forest and farmlands. Community groups representing agricultural and ecological interests in the area have thwarted efforts to build subdivisions and destination resorts. Opposition to development is intergenerational and deeply personal to many long-term residents. A local farmer recalled the beginning of their vigilance against development in the 1970s:

> [A developer from Portland] came into [my mom's] kitchen with a map, and the map showed that her farm was going to be like the twelfth hole of his golf course on his destination resort. And since that time, our family and our group that my mom helped form . . . has fought destination resorts on Mt. Hood, and so far, there's been like four proposals and we've basically won all of them.[25]

Despite the local opposition to development, some subdivisions have expanded on Hood River's west side. While this may help alleviate rising real estate prices closer to the city center, this development also lands on the Farmers Irrigation District's water rights. For that reason, the irrigation district has become responsible for municipal-style water utility services while also providing normal services to local farmers in the valley. Even though (as in the case of Bend) the residential water customers are not actually increasing the total water supply needed to serve the region, they have customer service expectations that differ from what irrigation districts are used to. In many cases they expect the kind of immediate service one would receive in a city; what they find is an understaffed irrigation district tasked with managing irrigation water for the agricultural industry. As one water manager put it:

> We've got a lot of urbanization in our district . . . so we're seeing subdivisions that are basically dividing up land that has water rights tied to it, and it's about half of our district that's non-agricultural use now . . . so it's a whole different animal to serve water to those folks.[26]

While the Farmers Irrigation District has historically served agricultural customers, half of their users are now residential, creating an uneven situation in which the farmers are outnumbered even though they still require higher quantities of water than residential subdivisions. At the same time, the residential areas require more attention from the irrigation district, generating frustration over the ignorance of new residents regarding water infrastructure. A water manager described getting "phone call after phone call after phone call" from residential customers unhappy with their water service. The frustration becomes a cultural struggle in which the newcomers are viewed as urbanites who do not understand rural places. As a water manager explained:

> Seventy percent of our calls come from residential users, and probably 70 percent of our staff time is spent helping people find their valves and helping them figure out how to use their water. You look at the demographics of the people moving here, and they're from Seattle or the Bay Area. They've never had irrigation

> water before. They have no idea how it works. . . . It's a pretty big customer base, and we're a small crew.[27]

The distinction between municipal water and irrigation water lies in questions of distribution, and farmers are tasked with creating their own infrastructure beyond the point to which the irrigation district delivers the water. However, some residents have complained that the new water users were being patronized and neglected by understaffed district managers. Additionally, many new water users find that their individual water rights under irrigation district regulations fall well short of supporting the kind of nicely manicured lawns, gardens, and landscaping that they expected from a rural lifestyle. One resident explained the tension:

> I kind of resented some of the way they look at some of the residential folks . . . the mere fact that they used the word *subsidizing.* [They say] they've been "subsidizing" the residential people, which in fact, the residential people are subsidizing Farmers [Irrigation District], because there are a lot more [residential] people paying fees.[28]

The difficulties of extended urbanization in the Farmers Irrigation District suggest that rural areas do not always easily adapt to the demands of amenity migrants in ways that accommodate residential development. At the same time, the residents find their vision of a tranquil and relaxed exurban lifestyle clashing with the reality of traditionally agricultural regions.

PIPELINE POLITICS: DEBATES AROUND IRRIGATION EFFICIENCY AND SENSE OF PLACE

Even though urban development itself does not necessarily stress regional water resources, farmers and irrigation districts in both the Deschutes River Basin and the Hood River Valley still must adapt to water resource pressures—in particular, the larger instream flow requirements associated with new endangered species listings. One key strategy being employed in both areas is modernization of old irrigation canal infrastructure through canal piping.[29] Much of the region's infrastructure built to convey agricultural irrigation water involves old and leaky irrigation canals. While some canals are lined with concrete to reduce water loss through seepage, many of the canals are unlined, and even lined canals leak if they are not constantly maintained. One of the main strategies for water conservation involves piping—that is, replacing the leaky and unlined canals with large buried pipes that are far more efficient at conveying water. Yet canal piping can be surprisingly contentious.

Hood River's irrigation districts are committed to piping canals in order to conserve valuable water resources. While most people understand this procedure as costly but effective and urgent, particularly given the impending stressors of climate change, some find it difficult to refrain from a degree of sentimentality. One local geographer recalled formative experiences with an irrigation ditch that proved a pedagogical and recreational resource for children and a source of employment:

> Years ago, I worked at a summer camp called Camp Baldwin, and there was a ditch there called Wolf Run Ditch. . . . It just came roaring through this Ponderosa forest, and it had over time gotten all kinds of ferns. . . . I loved it, we had little footbridges over it. About ten years after I was working at that camp, I was shocked, because I was walking around and was like, "Where'd the stream go?" And they had put it in a pipe and had it covered up, and it had enough pine needles on top that I couldn't find it . . . and it was so freaky, because this was a stream. . . . And I felt a little sentimental about it.[30]

In the arid Deschutes River Basin, feelings over canals and ditches have also proven contentious—according to some, even more so than in Hood River. Since the early 1900s, eight irrigation districts have provided water to over one hundred thousand acres of agricultural land within the river basin. The agricultural diversions have long altered the river's flows. In the winter, water is held back to recharge irrigation reservoirs, and in the summer, more water is released to travel downstream into irrigation canals. But the volcanic soils in the region are highly porous, and many of the canals leak large quantities of water. Hydrologists have estimated that 46 percent of the total water diverted from the Upper Deschutes River is lost through leaky canal bottoms.[31] Irrigation districts have been working to figure out how to conserve enough water to meet the needs of the endangered spotted frog while still continuing to support agriculture, and in October 2019, the irrigation districts jointly released a draft Habitat Conservation Plan detailing how the target flows would be met. Efforts to secure more water for spotted frog habitat have focused largely on addressing the inefficient irrigation canals.

Yet, while irrigation districts and other water managers in the Deschutes River area see canal piping as an important infrastructure project that will allow agricultural water use to continue while also leaving enough water in the streams for endangered species, many of the canals run through residential areas of Bend. In addition to moving irrigation water, the canals are seen as providing benefits of scenic beauty and increasing property values for homes with canal-front property. As piping projects have moved forward, they have been met with resistance from neighbors who do not want to lose their views and amenity landscapes. For example, as one water manager explained:

> In this area here around Bend and Redmond, there are people that are opposed to having the canal piped, more out of their own interests versus the interest of the greater good. . . . The gist of it is, people bought property with the canal running through it, and it's an amenity that was there when they purchased the property, but it's really a utility, just like a power pole.[32]

Another environmental advocate shared a story that illustrates the misunderstanding new residents can have regarding the canal infrastructure:

> I got a call from somebody wanting to know why "her river" was shut off. It turned out that she had just moved in during the summer and bought a house on one of the canals. And when we turned the canals off in the winter, "her" water went away and she was upset about it. She probably paid a premium price for the property and didn't realize that it really wasn't a river, it was a canal.[33]

Bend residents have made canal piping particularly difficult for irrigation districts by seeking official designation of canals on the National Register of Historic Places. In some cases, they have successfully obtained such historical designation, which makes it much more difficult for irrigation districts to make substantive changes to the canals (such as piping). Residents have made the case that the canals should receive historical designation because the canals represent important elements of the region's history that led to increased agricultural production and economic development in the rural area. However, irrigation districts have pushed back, noting that the segments of canal for which residents have sought historical designation are not necessarily the most representative or oldest historical infrastructure; instead, they tend to be conveniently located near residential areas.

This case of "pipeline politics" raises some interesting and unexpected tensions and alliances. The phenomenon of newly arrived residents, relocated from more densely urbanized locales, who oppose the canal piping projects for reasons of aesthetics and property values resonates with exurban political ecology literature that examines the politics of aesthetic landscape consumption and value conflicts.[34] However, the Deschutes situation is also somewhat unique. Given the pressures of the federal Endangered Species Act, the ability of Deschutes Basin farmers to continue agricultural production hinges on successfully leaving enough water in the river for the spotted frog while also preserving the ability to withdraw irrigation water—making conservation and water-efficiency projects the only viable option. Thus, agricultural irrigation districts have been advocating for conservation projects that will leave more water in rivers, whereas urban populations often assumed to be "environmentalists" push back against these same projects.

COLLABORATION AND ALLIANCES

While urbanization clearly presents a range of challenges for land and water governance, it is important to mention that urban populations can also introduce new opportunities and collaborative alliances.[35] New residents bring not only a range of new values, needs, politics, and conflicts but also new resources, ideas, and connections. As an environmental advocate described:

> There's a value shift for sure. This whole area, central Oregon, was built on irrigation and forestry and resource extraction. And that's

> shifted a lot, and it continues to shift as people move to Bend more for the river and recreation. . . . I think that that's creating cultural shifts, and [irrigation districts] probably feel more pressure than they used to feel to put water in the stream and accommodate growth and that kind of thing.[36]

In both Bend and Hood River, collaborative organizations have emerged to support river conservation, local agriculture, and urban communities, including working to strengthen community support for the much-needed canal piping projects. These collaborative organizations harness so-called newcomers' and old-timers' respect for healthy rivers along with desire to keep local agricultural economies working, channeling both into material support and funding. They are working effectively to develop shared visions and to build water governance models that work in practice for the various stakeholders in the basins.

CONCLUSIONS

In this chapter, we have examined urbanization processes in two smaller communities outside of Mainstreet Cascadia, focusing on the relationships between population growth and demographic change, on the one hand, and shifting relationships of water resource management, on the other. The changes in Bend and Hood River, Oregon, can be contextualized within the broader landscape of urbanization throughout the West Coast. Examining these cases helps develop an understanding of the multiple intersecting impacts of urbanization on water resources in communities outside the region's major cities.

Eastern Oregon is undergoing multiple changes simultaneously, including demographic change, such as population growth driven largely by broader regional-scale real estate prices and amenity migration; economic change, including shifts from timber toward growing recreation and tourism industries; and biophysical change, including shifting precipitation regimes associated with climate change.[37] Together, these changes have the potential to significantly impact water resource demand and management. However, as we describe in this chapter, these impacts are not what one might initially expect. Bend and Hood River are both experiencing increasing flows of new residents related to broader push factors of increasing housing prices in the major metropolitan areas of the West Coast as well as pull factors of ample outdoor recreation and rural amenities. Yet in both of these communities, despite rapid population growth, urban water demand is not the greatest pressure on water resources. Instead, irrigated agricultural water use continues to dominate water use, accounting for the vast majority of water rights. In Hood River, irrigation districts have not struggled to meet absolute water demands in terms of gallons needed but have instead struggled to cope with the customer service demands of new exurban residents on formerly agricultural lands. In the Bend area, efforts to conserve water to protect instream flows for endangered species have generated new forms of conflict, as efforts to pipe leaky irrigation canals have sparked tensions between irrigation districts and residents around what water is for and whether it is primarily an aesthetic amenity or an input for agricultural production. Meanwhile, in both Bend and Hood River, collaborative organizations have arisen, emphasizing common ground and seeking solutions that work for farmers, urban residents, and rivers.

We conclude with a reflection on the broader dynamics of urbanization and water use, which are not only regional in scale. There are larger processes of urban metabolism at work as well, through which water "circulates" not only within watersheds but around the world. The water used to irrigate rural Oregon's crops is not separate from global urbanization processes. This water enters into a globalized food system in the form of agricultural products such as pears and carrot seeds. Although rural residents and farmers may not often think of themselves as part of extended urbanization, farmers and water from Hood River and the Deschutes River Basin are networked into a global agricultural production and trade system that keeps supermarket shelves stocked worldwide, enabling further urban growth, which then overflows into the exurban dynamics described here. Thus, urbanization includes not only the migration of people from larger cities into smaller towns; agricultural activity in the Hood River Valley and the Deschutes River Basin can also be considered part of a broader urban metabolism.

From a perspective of urbanization and urban metabolism, water is an input into a global agricultural system, in which it circulates around the world in the form of agricultural products and produce. At the same time, as this chapter shows, water is also an important part of the landscape, driving conflicts over local residents' experiences of place. Water also provides a nexus for collaboration, as people with different histories and experiences come together over a shared vision of healthy rivers. This examination of Bend and Hood River, Oregon, demonstrates how flows of people, ideas, agricultural goods, and water coproduce their immediate regions as well as the broader region of Cascadia and beyond.

NOTES

We recognize that contemporary water management is rooted in settler colonialism. We acknowledge the Tenino people of the Deschutes River region, represented today by the Confederated Tribes of Warm Springs, along with the Wasco and Pauite people who were forced to relocate to the Warm Springs Reservation. We also acknowledge the Multnomah, Kathlamet, Clackamas, Chinook, Tualatin, Kalapuya, Molalla, and many other Tribes who made and make their home along the Columbia River.

1 Ash Amin and Nigel Thrift, *Cities: Reimagining the Urban* (Cambridge: Polity Press, 2002).

2 Raymond Williams, *The Country and the City* (New York: Oxford University Press, 1973); Hillary Angelo and David Wachsmuth, "Urbanizing Urban Political Ecology: A Critique of Methodological Cityism," *International Journal of Urban and Regional Research* 39, no. 1 (2015): 16–27; and Innisfree McKinnon, "Competing or Compatible Capitalisms? Exurban Sprawl and High-Value Agriculture in Southwestern Oregon," in *A Comparative Political Ecology of Exurbia: Planning, Environmental Management, and Landscape Change*, ed. Laura Taylor and Patrick T. Hurley (Cham, Switzerland: Springer, 2016), 105–30.

3 Thanks to Gail Snyder, executive director of the Coalition for the Deschutes, for her valuable insights and feedback. Thanks to Stephen Barro for helpful research assistance. Huge thanks to all the interviewees in the Bend and Hood River areas for sharing their time and perspectives.

4 We conducted a total of forty-two semi-structured interviews in Hood River and ten in Bend. Interviewees were chosen by snowball sample and were selected to represent a range of areas of expertise and perspectives on water management. All interviews were conducted confidentially, and names are withheld in this chapter to protect the privacy of interviewees. We also observed public meetings, including watershed coalition meetings, listening sessions and public forums around water issues, and field tours. We supplemented and triangulated information from interviews with news articles from the *Bend Bulletin* and the *Hood River News* and technical and policy reports.

5 Nik Heynen, Maria Kaika, and Erik Swyngedouw, *In the Nature of Cities: Urban Political Ecology and the Politics of Urban Metabolism* (London: Routledge, 2006).

6 William Cronon, *Nature's Metropolis: Chicago and the Great West* (New York: W. W. Norton, 1991); Richard A. Walker, *The Country in the City: The Greening of the San Francisco Bay Area* (Seattle: University of Washington Press, 2007).

7 Alida Cantor, "Hydrosocial Hinterlands: An Urban Political Ecology of Southern California's Hydrosocial Territory," *Environment and Planning E: Nature and Space* (March 2020), https://doi.org/10.1177/2514848620909384.

8 Neil Brenner, "Theses on Urbanization," *Public Culture* 25, no. 1 (2013): 85–114.

9 Colleen C. Hiner, "Beyond the Edge and in Between: (Re)conceptualizing the Rural-Urban Interface as Meaning-Model-Metaphor," *Professional Geographer* 68, no. 4 (2016): 520–32.

10 Taylor and Hurley, *Comparative Political Ecology.*

11 Taylor and Hurley, *Comparative Political Ecology.*

12 Paul Robbins, Katharine Meehan, Hannah Gosnell, and Susan J. Gilbertz, "Writing the New West: A Critical Review," *Rural Sociology* 74, no. 3 (2009): 356–82; and Joseph E. Taylor III, "The Many Lives of the New West," *Western Historical Quarterly* 35, no. 2 (2004): 141–65.

13 Jesse Abrams, "Rural Residential Development and Its Discontents: A Political Ecology of Sprawl Containment in Wallowa County, Oregon, USA," in Taylor and Hurley, *Comparative Political Ecology*, 83–103.

14 McKinnon, "Competing or Compatible Capitalisms?"

15 Colleen C. Hiner, "'Chicken Wars,' Water Fights, and Other Contested Ecologies along the Rural-Urban Interface in California's Sierra Nevada Foothills," *Journal of Political Ecology* 23, no. 1 (2016): 167–81.

16 Meerah Powell, "Report: 40,000 Californians Move to Oregon Annually," *Portland Tribune*, May 5, 2019, https://pamplinmedia.com/pt/9-news/427488-334551-report-40000-californians-move-to-oregon-annually-.

17 The US Census Bureau identifies two types of urban areas: urbanized areas of 50,000 or more people and urban clusters of at least 2,500 and less than 50,000 people. Rural is defined by the Census as all population and territory not included within these urban areas. See www.census.gov/programs-surveys/geography/guidance/geo-areas/urban-rural.html. Taylor and Hurley, *Comparative Political Ecology.*

18 Oregon Environmental Council, *Making Water Work: Strategies for Advancing Water Conservation in Oregon Agriculture* (Portland: Oregon Environmental Council, 2012), https://oeconline.org/wp-content/uploads/2014/12/Making-Water-Work_web.pdf; and Jeffrey Mount and Ellen Hanak, "Water Use in California," Public Policy Institute of

California, May 2019, www.ppic.org/publication/water-use-in-california/.

19 Brent Olson, "'In the Real Estate Business Whether We Admit It or Not': Timber and Exurban Development in Central Oregon," in Taylor and Hurley, *Comparative Political Ecology*, 131–45.

20 Gavin Bridge, "Material Worlds: Natural Resources, Resource Geography and the Material Economy," *Geography Compass* 3, no. 3 (2009): 1217–44.

21 Deschutes water manager, interview by the authors, September 24, 2019.

22 Deschutes environmental advocate, interview by the authors, September 11, 2019.

23 An acre-foot is a measurement of water equal to about 326,000 gallons, or enough water to cover one acre of land one foot deep.

24 Deschutes water manager, interview by the authors, September 24, 2019.

25 Hood River farmer, interview by the authors, October 22, 2019.

26 Hood River water manager, interview by the authors, October 8, 2019.

27 Hood River water manager, interview by the authors, October 8, 2019.

28 Hood River local resident, interview by the authors, November 25, 2019.

29 Headwaters Economics, *Agriculture and Irrigation in Oregon's Deschutes and Jefferson Counties* (Bozeman, MT: Headwater Economics, 2017), https://headwaterseconomics.org/wp-content/uploads/Deschutes_River_Basin_Agricultural_Report.pdf.

30 Hood River geographer, interview by the authors, November 5, 2019.

31 Marshall W. Gannett, Kenneth E. Lite Jr., David S. Morgan, and Charles A. Collins, *Groundwater Hydrology of the Upper Deschutes Basin, Oregon: USGS Water Resources Investigations Report 00-4162* (Portland: US Department of Interior, US Geological Survey, 2001), https://pubs.usgs.gov/wri/wri004162/.

32 Deschutes water manager, interview by the authors, September 12, 2019.

33 Deschutes environmental advocate, interview by the authors, September 13, 2019.

34 Hiner, "Chicken Wars."

35 McKinnon, "Competing or Compatible Capitalisms?"

36 Deschutes environmental advocate, interview by the authors, September 11, 2019.

37 While real estate prices in both Bend and Hood River are still well below those in regional metropolitan areas such as Seattle and the California Bay Area, prices have increased dramatically in recent years to the point where newer and younger residents are now seeking alternatives. For example, Prineville and Redmond, two smaller towns near Bend, have been seeing more growth as Bend gets more expensive.

PART 2

INEQUALITIES

CHAPTER 5

TALES OF THREE CITIES

Urban History, Settler Colonialism, and Indigenous Survivance in Seattle, Vancouver, and Victoria

COLL THRUSH

IN THE WINTER OF 1919 AND 1920, SEATTLE'S L. C. SMITH Building was the tallest structure west of the Mississippi. Named for its builder, the industrialist and firearms manufacturer Lyman Cornelius Smith, it was in many ways a symbol of Seattle's arrival as a modern American city. Gleaming white, nearly five hundred feet tall, with a pyramidal top, and featuring a room furnished by the Chinese empress Cixi, the Smith Tower looked out over Elliott Bay and Puget Sound and represented Seattle's economic and cultural connections to the Pacific and beyond. It was also a marker of the city's Progressive Era optimism. It remains a key element of Seattle's twenty-first-century skyline, standing at the south edge of downtown's ever-expanding forest of high-rises and looming over the city's historic Pioneer Square and International District neighborhoods. The Smith Tower is iconic; indeed, the tallest building in the city, 1985's jet-black Columbia Center, is colloquially referred to as "the box the Smith Tower came in."[1]

In the winter of 1919 to 1920, the Smith Tower also looked across Elliott Bay to the mouth of the Duwamish River, which had been utterly transformed in the first years of the twentieth century. Flowing down out of the Cascade Mountains, the Duwamish gathered waters from the Green, Black, Cedar, and White Rivers, but by the time the Smith Tower was built, only the Green remained; the White and Cedar had both been diverted, and the Black had ceased to exist. Its lower channel straightened, an artificial island built in its estuary, and its banks crowded with industrial development, the Duwamish barely resembled its pre-urban self, when the Indigenous people who gave it its name were the only residents of its lower valley.

Some of those Duwamish people remained along the river. A man named George Seetoowathl and his unnamed wife, for example, lived in a floathouse on the lower Duwamish, eking out a living among the factories and docks. They, like many of their Duwamish relations, continued to hold knowledge of the land and waters as they were before the rise of the city named after a Duwamish headman. At least some of this knowledge was recorded by an anthropologist, including the many Indigenous place-names along the river and the shores of Elliott Bay. But in the winter of 1919 to 1920, the city's rapid development, which included profound environmental changes, proved too much for Seetoowathl and his spouse. That winter, the anthropologist reported, the couple starved to death on the river named for their people.[2]

That Indigenous people could die in such dire circumstances within sight of a growing metropolis, with all its rhetoric of opportunity and growth, speaks to a central truth of the region's history: throughout Coast Salish territories, the rise of cities such as Seattle was one of the primary fronts of settler colonialism and Indigenous dispossession.[3] As urban development utterly transformed the natural landscapes of the region, Indigenous people such as the Duwamish were often pushed aside and marginalized, with consequences that would last for decades. And Seattle was not unique; as this chapter argues, despite important differences, similar events were taking place in other parts of urban Cascadia, such as Vancouver and Victoria. Famously described as a "change of worlds" by Duwamish headman Seeathl, more widely known as Chief Seattle, the story of American, British, and Canadian settler colonialism in the Coast Salish world is one of profound and traumatic transformation.

Indeed, the stories of these three cities, when considered alongside one another, illustrate the ways in which urban development has been a key, if often underexamined, element of the region's settler colonial pasts and presents, a kind of civic imperialism that has had devastating environmental and societal consequences. But lest we write Indigenous people out of Cascadia's urban history, it is important to remember that despite the transformations of the landscape and other attempts to remove Indigenous communities from the homelands they had inhabited since time immemorial, those communities have maintained connections to their traditional territories through cultural memory, political and legal activism, and simple physical presence (see chapter 6 and map I.2 in introduction). Indigenous survivance, as we shall see, is also part of the urban story, and the Indigenous histories of Seattle, Vancouver, and Victoria, like settler colonialism itself, are ongoing. A century on from that deadly winter, these histories are far from over.

BUILDING CITIES IN COAST SALISH TERRITORIES: URBAN DEVELOPMENT AND INDIGENOUS MARGINALIZATION

The great irony of the history of urbanization in Cascadia is that Indigenous people and peoples were a crucially important part in the very founding and early development of places such as Victoria, Vancouver, and Seattle. Despite the commonly held notion that Indigenous histories have little to do with their urban counterparts, around the shores of the Salish Sea, these two kinds of history are deeply entangled with each other. It was not only the case that all three cities were built in Coast Salish territories; it was also the case that Coast Salish and other Indigenous people were central to the founding and early development of each urban center.

In Victoria, for example, founded around a fur trade fort in 1843, both local and recently arrived Indigenous people played a key role in the settlement's early success. Many of the prominent Hudson's Bay Company families were of mixed ancestry, and Indigenous women such as Amelia Douglas (Cree), Josette Legacé Work (Métis), Isabella Ross (Ojibwe), Mathilda McNeil (Kaigani Haida), Neshaki McNeil (Nisga'a), and Sophia Lolo (Secwepmc) counted themselves among the town's elite.[4] Meanwhile, the Songhees people, in whose territory the outpost had been founded, continued to live in their village directly across the harbor from Victoria proper, in a line of longhouses described by one visitor in 1846 as being "arranged with some degree of order in streets and lanes with passages running up between them."[5] And as the fur trade and other economic ventures began to suture the region together, Indigenous people from farther north along the coast made Victoria part of their annual rounds, sometimes outnumbering white residents. The story of Victoria's early development, then, cannot be told without placing Indigenous people at its center.

In Seattle, founded in 1852, a similar story emerges. Indigenous people from the Duwamish and other local communities were crucially important in the "village period" that stretched into the 1870s. They helped clear land and build houses; they worked in homes and in the sawmill; they provided food and carried the mail; and in a host of other ways, they facilitated the growth of the settlement from a precarious experiment on Elliott Bay to a burgeoning town that would go on to dominate the region.[6] Similarly, in Granville, the timber town that would ultimately be renamed Vancouver, local Indigenous nations such as the Sk̲wx̲wú7mesh (Squamish), səl̓ílwətaʔɬ (Tsleil-Waututh), and xʷməθkʷəy̓əm (Musqueam) would play a key role in the town's early development, in much the same way that their Coast Salish kin had done in Seattle and Victoria.[7]

But in all three places, trouble was brewing. In the latter decades of the nineteenth century, the trio of towns saw campaigns to limit Indigenous presence in the name of "progress." In Seattle, some of the earliest civic ordinances had forbidden Indigenous people from residing within the town limits unless they were employed and given shelter by white citizens—highlighting both the racism of settlers but also their need for Indigenous labor. Other laws governed the public presence and movement of Indigenous women, and the territorial legislature placed draconian strictures on the civil and political rights of mixed-race people, who made up a significant part of the population. Meanwhile, the importation of white women on "bride ships" was designed to "civilize" a place that had been fundamentally shaped by the presence and labor of Indigenous people. Combined with accelerating changes to the physical environment and the deliberate destruction of Duwamish villages in and around the growing city, such processes placed intense pressure on local Indigenous populations.[8]

In Victoria, increasingly virulent racism toward the mixed-heritage founding families meant that some of their progeny were gradually pushed aside in favor of white British families. While the daughters of these founding families often remained part of elite colonial society, including receiving education in Britain and marrying white men, the sons were more likely to be categorized as "Indian" and to marry Indigenous women, relegating them at best to the edges of polite society. Gender, it appeared, worked in important if unexpected ways in the capital of the colony. Bride ships were part of Victoria's history as well, further confirming the fact that ideas about white womanhood were central to the project of settler colonialism.[9] Meanwhile, ecological factors, most notably an outbreak of smallpox that spread from Victoria up and down the coast in 1862, transformed Indigenous life at the southernmost tip of Vancouver Island and beyond.[10] And in Vancouver, attempts to tame Indigenous sexuality through policing, the establishment of Indian reserves, the condemnation of longhouses outside those reserves, and the destruction of dozens of salmon streams all conspired to make life very difficult indeed for Indigenous people living in and around the fast-growing town.[11] At the turn of the twentieth century, it seemed to many observers that Indigenous history had ended not long after urban history had begun.

But it was in the first years of the twentieth century that the final blows seemed to fall. The Progressive Era, with its emphases on orderly management of the environment, the aesthetics of hygiene, and a vision of implacable modernity, proved especially challenging for Indigenous people living in and around the three cities. To be sure, there were important distinctions between Vancouver, Seattle, and Victoria. The territory claimed by Vancouver included two Indian reserves (and several more nearby), while there was no reservation in or close to Seattle—white settlers had made sure of that in the 1860s with a petition signed by nearly every white man in King County. At the same time, treaties had been the law of the land in Washington State since 1855 (even if most treaty rights were not acknowledged or respected in meaningful ways until the 1970s), whereas most of British Columbia remained unceded. To complicate things further, parts of Victoria were covered by a handful of small treaties signed in the early 1850s. Across the Coast Salish world, the exigencies of settler colonial politics created a diverse and often confounding legal landscape that settlers and Indigenous people alike struggled to navigate.

But for all their important differences, the three cities' urban and Indigenous histories seemed to align in crucial ways in the early twentieth century. In Vancouver, for example, the reserve at Kitsilano, which included the ancestral community of Seńáḵw, was described by one visitor as "a settlement of consequence,"[12] but in the eyes of many white city leaders and residents, the continued presence of Seńáḵw was an impediment to progress. Beginning in the first years of the twentieth century, an alliance of businessmen, federal authorities, and even Christian organizations began to pressure the residents of Seńáḵw, which many observers perceived as an unproductive and unhygienic eyesore, to leave. The Squamish inhabitants resisted. During a 1904 meeting on the reserve, Chief Chip-kay-um "didn't want to leave this place where he was born and it is the place where his dead relatives are buried," according to one Indian agent. "None of the men on the place," he continued, "want to sell it—the Queen gave him and his people the land."[13] Finally, though, in 1913, the people of Seńáḵw were paid a small sum each, loaded on a barge, and sent away to a reserve across Burrard Inlet, their longhouses set afire as the barge pulled away from shore. The forces of civic progress, it seemed, had prevailed.

A similar process had already taken place in Victoria, where the Songhees village was the target of virtually the same discourses of urban order. Such pressures had their origins in the nineteenth century. In 1890, one Victoria resident reported that the Songhees were "to a man determined not to vacate the . . . reserve which was held and occupied by their forefathers from time immemorial and consequently dearer to them than anything they could possess."[14] As historian Jean Barman notes, "There was no reason for the Songhees to want to move, and, in fact, they did not want to do so."[15] However, mounting pressure, along with financial enticements, opened fissures within the Songhees community. Historian Grant Keddie, for instance, has noted that while some residents were in favor of leaving, others were firm on the land question.[16] The end result, however, was much like what would happen at Seńáḵw: in 1911, the Songhees were removed to a more distant reserve, opening the shoreline to urban development. As historian Penelope Edmonds has noted, the expulsion of the Songhees from Victoria signified larger and more fundamental transformations in the local economy as it transitioned from a mercantilist system centered on the fur trade to a robust settler-colonial economy that had little place for preexisting Indigenous systems of governance, land tenure, or livelihood.[17]

If 1911 and 1913 were especially important years in the Indigenous histories of Victoria and Vancouver, respectively, that same decade saw parallel developments taking place in Seattle. Most notably, the completion of the Hiram M. Chittenden Locks in 1916, which connected Lake Washington and Lake Union with Puget Sound, would be particularly devastating for Duwamish people. The locks, justified as a rational "improvement" on nature, forced out Indigenous people who were still living on the site of their construction, just a stone's throw from the ancestral community of Tucked Away Inside. Meanwhile, the opening of the locks completely reoriented the city's watersheds and lowered Lake Washington by several meters, meaning that the Black River south of the city, where many Duwamish people had come to reside, ceased to exist in the course of a single afternoon. Combined with the straightening and industrialization of the Duwamish River's lower reaches, the last years of the decade were disastrous for local Indigenous people.[18]

Taken together, the stories of the three cities bear strong resemblances to one another, in the sense that they are part of a much larger regional story of urban dispossession in Coast Salish places. That the rise of new towns and cities in Cascadia has been one of the vanguards of settler colonialism in the region should come as no surprise; it is in the nature of urban places to utterly transform landscapes and the lives lived upon them. For the people of Tucked Away Inside and Seńákw and Songhees, the births of Seattle, Vancouver, and Victoria, respectively, did indeed spell the end of some things, most notably ways of life that drew direct bodily sustenance from land and water. However, ending the story in the 1910s would offer an incorrect picture of both urban and Indigenous history. Ending with the liquidation of Songhees and Seńákw, and the locks and other changes in Duwamish territory, leaves out a critically important element of Cascadian history: Indigenous survivance.

URBAN SURVIVANCE AND MUNICIPAL COLONIALISM: INDIGENOUS PERSISTENCE IN URBAN PLACES

"The condition of Indigeneity in North America," writes the Mohawk anthropologist and leading Indigenous studies scholar Audra Simpson, "is to have survived."[19] And so, despite the seemingly total dispossession of Indigenous peoples from Seattle, Vancouver, and Victoria in the early twentieth century, a century on, those peoples are still present in the three cities. They are perfect examples of what

Ojibwe scholar Gerald Vizenor has called survivance: the idea that Indigenous people have continued to exist into the present and, more than simply surviving, have resisted, adapted, and even in some cases thrived, despite the near-apocalyptic horrors of settler colonialism.[20] It is to that story of survivance, built into the very fabric of urban Cascadia, that we now turn.

The Progressive Era is long gone, and the three cities at the heart of our story are fundamentally different places than they were a century ago. Seattle has become world famous for its contributions to the corporate, cultural, and even political worlds. Home to Amazon and Starbucks, the city also gave us grunge and the World Trade Organization protests of 1999. Vancouver is an archetypal Pacific Rim city, with a very large Asian-descent population and high real estate values; it seemed to emerge onto the world stage both with the World's Fair of 1986 and the 2010 Winter Olympics. And Victoria, with its slower island pace, has cast its lot with Anglophilia, crafting a quaint English identity for itself and becoming popular with Canadian retirees. All three cities, meanwhile, have used Indigenous symbolism to sell themselves, most notably in the totem poles, new arrivals from the north, that serve as urban icons.

The presence of Tlingit or Kwakwa̱ka̱'wakw poles in Duwamish, Squamish, Musqueam, Tsleil-Waututh, and Songhees territories—places where totem poles per se had not traditionally existed—might easily overwrite the history of these local peoples and nations. Indeed, in many cases, the imposition of northern-style totem poles emerged out of the cities' claims to the broader northwest coast. In the case of Seattle, its first totem pole was erected in 1899 after being stolen from the Tlingit community in Alaska, serving as a marker of the American city's experience with the boom of the Klondike Gold Rush and thus a symbol of urban empire. In both Vancouver and Victoria, the prominent display of North Coast–style poles was always linked to civic branding and urban spectacle.

And indeed, in the twentieth century, urban empire continued across Coast Salish territories. In Vancouver, in the middle third of the century, city leaders looked back on Victoria's dispossession of the Songhees as a model for what historian Jordan Stanger-Ross has called "municipal colonialism." While local First Nations resisted land loss, invoking their unique relationship with the Crown over the prerogatives of city leaders, the reality was that the lands they still held in reserve were under constant threat, whether from largely unchecked industrial development or the expansive creep of settler neighborhoods.[21] Meanwhile, families of Coast Salish and other ancestries continuing to live in Vancouver's iconic Stanley Park were evicted in favor of an unpeopled "nature," while archaeological investigations in the city often ignored local Indigenous communities, instead imagining a "mysterious people" that had once inhabited the area before the local people who were still there.[22]

In Seattle, Duwamish and other Indigenous fishers saw their harvesting practices criminalized for much of the twentieth century, despite the fact that it was continued habitat loss thanks to urban and suburban development, not Indigenous fishing, that was truly decimating salmon populations. It was only in the early 1970s that treaty fishing rights were finally upheld by the courts, but soon after, the Duwamish lost their status as a federally recognized tribe, a devastating decision that hinged, in part, on the dislocations of the early twentieth century.[23] And while Victoria's midcentury Indigenous history is less well documented than that of the other two cities, the removal of the Songhees from the inner harbor gave the impression that Indigenous people had never been there, at the same time that totem poles at the Royal British Columbia Museum and elsewhere in town provided Victoria with a sense of regionally grounded identity that served as a counterpoint to the "Ye Olde England" character city leaders had capitalized on to draw tourists.

Despite these ongoing forms of settler colonialism, in all three cities, Indigenous people maintained connections to traditional places within the urban fabric. In Vancouver, Squamish elder August Jack Khatsalano, who had lived at Seńáḵw, held long conversations with historian Major Matthews from the 1930s to the 1950s, offering drawings of the longhouses, maps of nearby place-names, descriptions of a nineteenth-century potlatch, and more, documenting in concrete ways the continuation of Indigenous knowledge pertaining to the civic landscape.[24] Decades later, Tsleil-Waututh writer and scholar Lee Maracle wrote elegiacally of Seńáḵw's place in local Indigenous memory: "Raven has never left this place. Sometimes it feels like she has been negligent, perhaps she fell asleep and never woke up. But Raven has not left this place."[25] Meanwhile, descendants of the families ousted from Stanley Park held onto stories of their relatives' homes and lives, ultimately sharing them with historian Jean Barman.[26]

In Seattle, the Duwamish, despite governmental and bureaucratic challenges to their status as a tribe, maintained a sense of themselves as a people through kinship networks and political activism that included everything from protests when archaeological sites were disturbed by urban development to the dogged pursuit of federal recognition in the courts. When the city celebrated its sesquicentennial in 2001, the Duwamish were on hand to remind Seattle's settler population that the Duwamish were still there.[27] And in Victoria, where Songhees presence was perhaps more muted, the fact that the provincial legislature was located on the inner harbor meant that Songhees territory remained a site of activism and advocacy for Indigenous people from around British Columbia. Lastly, in all three cities, growing urban Indigenous populations, with roots often far beyond the local communities in whose territories the cities had been built, meant that Vancouver, Victoria, and Seattle remained Indigenous spaces.

And so today, a century on from the dispossessions of the Progressive Era, it is impossible to move through any of Cascadia's major cities without seeing evidence of the first peoples of those places. In Vancouver, the 2012 disturbance of ancestral burials at a four-thousand-year-old village site inspired the Musqueam and their allies to hold a two-hundred-day vigil at the location, resulting finally in the preservation of the place called c̓əsnaʔəm.[28] Meanwhile, Musqueam artist Quelemia Sparrow created a site-specific performance piece entitled *Ashes on the Water* that told the story of white and Indigenous women coming together in the wake of the city's great fire of 1886, reminding audiences that Indigenous presence in the city had never ended and gesturing toward the role that the Musqueam and other local peoples had played in the city's early history.[29] And after a thirty-year legal battle, the Squamish First Nation has reasserted control over the reserve in Kitsilano, recently announcing plans for a group of high-rises to be built on the reserve land; the development is called, appropriately, Seńáḵw. In artists' renderings, the towers are marked with Coast Salish–style detailing and green roofs, speaking to the values of the Squamish.[30] Similar development projects are in the works around the city, including a new urban neighborhood at Jericho near the site of the ancestral village of ʔəy̓alməxʷ and a large housing project near the University of British Columbia simply named leləm̓, meaning "home" or "house."[31] Meanwhile, the City of Vancouver, like the province of British Columbia, has promised to uphold the United Nations Declaration on the Rights of Indigenous Peoples, recognizing all three local First Nations. (Whether either promise will amount to anything remains to be seen; at this writing, Indigenous-led blockades of ports and railways across Canada, a response to attempts to build pipelines through unceded Indigenous territories in northern British Columbia, are dominating the headlines.)

Entering the third decade of the twenty-first century, parallel developments in Victoria and Seattle similarly speak to the ongoing presence of Indigenous communities. In the British Columbia capital, the Songhees, after many years, reached a settlement with the province and the government of Canada over their failure to honor the 1850 treaty, and a series of art installations now commemorate the history of Songhees people on the inner harbor.[32] In Seattle, the Duwamish continue their efforts toward federal recognition, with widespread support from their non-Indigenous neighbors, manifested in such things as efforts to preserve a sacred spring in a city park and the establishment of a tithing-style campaign of financial support under the hashtag #RealRentDuwamish.[33] Meanwhile, area tribes such as the Muckleshoot and Suquamish, who have maintained federal recognition, still assert treaty-based authority over, for example, salmon populations in Elliott Bay and the Duwamish River.[34]

Together, these stories of continued Indigenous presence in urban spaces speak to the proof of survivance: the creative and compelling resistance to the forces of settler colonialism and the refusal to be totally, irrevocably removed. They challenge the idea, so foundational to much of the way North American history is narrated, that Indigenous and urban histories are somehow mutually exclusive. Instead, they illustrate the ways in which the two kinds of history are bound up with each other: mutually constitutive, entangled, imbricated. They also force us to consider the ways in which, despite its profound effects, the urban development of Cascadia is but a recent and perhaps even ephemeral phenomenon when compared to the long inhabitance of Indigenous peoples in these places. The settler histories of Seattle, Victoria, and Vancouver are in fact temporally shallow, beginning only yesterday in the grand scheme of things. A little over a century and half is, in truth, a very short time indeed.

URBAN CASCADIA AS INDIGENOUS HISTORY: ASSESSING THE CHANGE OF WORLDS

Every city in Cascadia has Indigenous history. Auburn and Abbotsford, Olympia and Nanaimo, Portland and Powell River, and all the rest have played a central role in the story of settler colonialism in the region. (Danica Miller's contribution to this book, for example, sets the urban history of Tacoma against her own Puyallup people's much longer history in place.) Most of these histories have not been written down, but if I were to hazard a guess as to what one might find in the colonial and Indigenous archives of these other places, I would expect the following: the meaningful presence of Indigenous workers and relatives in the settlements' early histories, the gradual—and likely at times sudden—dispossession from ancestral spaces, the persistence of Indigenous memory and sometimes practices in traditional places, and, especially in recent decades, forms of activism that put the lie to the notion that Indigenous peoples have somehow disappeared.

In the end, all urban history is Indigenous history, because every urban space in Cascadia—all the way down to a single bocce ball park in Vancouver[35]—is built on Indigenous land, treatied or not. Seetoowathl and his wife, who died on the Duwamish a century ago, are but one example of the human costs of urban development in Coast Salish territories, unceded or otherwise. But to focus only on their demise, brought into sharper focus by their nearness to the urban spectacle of the Smith Tower, can lead us to think that Indigenous history was always already ending as soon as urban history began. This is nothing more than colonial fantasy. The truth is that the Duwamish, Musqueam, Tsleil-Waututh, Squamish, and Songhees people continue to maintain relationships with their ancestral and contemporary landscapes. They are not going anywhere, and, of course, neither are the rest of us. The way forward, then, is to learn how to live together in the debris of settler colonialism, in the wake of a change of worlds.

NOTES

Coll Thrush lives and works in the traditional, ancestral, and unceded territories of the Musqueam, Squamish, and Tsleil-Waututh Nations, currently known as Vancouver, British Columbia. For the relationship between the University of British Columbia's Point Grey campus and the Musqueam First Nation, visit https://indigenous.ubc.ca/indigenous-engagement/musqueam-and-ubc/.

1 For the history of Smith Tower, see John Pastier, "Smith Tower (Seattle)," HistoryLink.org, July 1, 2004, www.historylink.org/File/4310.

2 Coll Thrush, *Native Seattle: Histories from the Crossing-Over Place*, 2nd ed. (Seattle: University of Washington Press, 2017), 79, 103.

3 "Coast Salish" was originally an anthropological category based on linguistics and shared cultural practices, referring to Indigenous peoples in parts of Washington and British Columbia. In recent years, it has become more common in Indigenous circles as a framing that acknowledges long-standing historical connections between the diverse communities of the region.

4 Sylvia Van Kirk, "Tracing the Fortunes of Five Founding Families of Victoria," *BC Studies*, no. 115/116 (1997/1998): 149–80.

5 Berthold Seemann, *Seemann's Voyage of the* Herald, vol. 1 (London: Reeve & Co., 1853), quoted in Grant Keddie, *Songhees Pictorial: A History of the Songhees People as Seen by Outsiders, 1790–1912* (Victoria: Royal British Columbia Museum, 2003), 24, 26.

6 Thrush, *Native Seattle*, 17–65. See also David M. Buerge, *Chief Seattle and the Town That Took His Name: The Change of Worlds for the Native People and Settlers on Puget Sound* (Seattle: Sasquatch Books, 2017).

7 For the early history of Vancouver, see Robert A. J. McDonald, *Making Vancouver: Class, Status, and Social Boundaries, 1863–1913* (Vancouver: UBC Press, 1996).

8 Thrush, *Native Seattle*, 40–78.

9 Van Kirk, "Tracing the Fortunes."

10 For a detailed account of the 1862 outbreak, see Greg Lange, "Smallpox Epidemic of 1862 among Northwest Coast and Puget Sound Indians," HistoryLink.org, February 4, 2003, www.historylink.org/File/5171. For the larger context of epidemic diseases in the region, see Robert Boyd, *The Coming of the Spirit of Pestilence: Introduced Infectious Diseases and Population Decline among Northwest Indians, 1774–1874* (Seattle: University of Washington Press, 1999).

11 Jean Barman, "Taming Aboriginal Sexuality: Gender, Power, and Race in British Columbia, 1850–1900," *BC Studies*, no. 115/116 (1997/1998): 237–66.

12 Rev. C. M. Tate in conversation with Major J. S. Matthews, July 1, 1932, in J. S. Matthews, *Conversations with Khahtsahlano, 1932–1954* (Vancouver: Vancouver City Hall, 1955), 158–59, 172. Quoted in Jean Barman, "Erasing Indigenous Indigeneity in Vancouver," *BC Studies*, no. 155 (2007): 8–9.

13 R. C. McDonald, Indian agent, cited in Barman, "Erasing Indigenous Indigeneity," 14.

14 A. W. Vowell to L. Vankoughnet, deputy superintendent of Indian affairs, May 5, 1890, Department of Indian Affairs (DIA), RG 10, vol. 3688, file 13.886-1. Quoted in Barman, "Erasing Indigenous Indigeneity," 6.

15 Barman, "Erasing Indigenous Indigeneity," 6.

16 Keddie, *Songhees Pictorial*, 146–54.

17 Penelope Edmonds, "Unpacking Settler Colonialism's Urban Strategies: Indigenous Peoples in Victoria, British Columbia, and the Transition to a Settler-Colonial City," *Urban History Review* 38, no. 2 (2010): 4–20.

18 For an overview of Seattle's environmental history and its effects on Indigenous people, see Coll Thrush, "City of the Changers: Indigenous People and the Transformation of Seattle's Watersheds," *Pacific Historical Review* 75, no. 1 (2006): 89–117; Matthew Klingle, *Emerald City: An Environmental History of Seattle* (New Haven, CT: Yale University Press, 2007); and David B. Williams, *Too High and Too Steep: Reshaping Seattle's Topography* (Seattle: University of Washington Press, 2015).

19 Audra Simpson, "Settlement's Secret," *Cultural Anthropology* 26, no. 2 (2011): 205–17.

20 See, for example, Gerald Vizenor, ed., *Survivance: Narratives of Native Presence* (Lincoln: University of Nebraska Press, 2008).

21 Jordan Stanger-Ross, "Municipal Colonialism in Vancouver: City Planning and the Conflict over Indian Reserves, 1928–1950s," *Canadian Historical Review* 89, no. 4 (2008): 541–80. See also Nicholas Blomley, *Unsettling the City: Urban Land and the Politics of Property* (New York: Routledge, 2004).

22 Jean Barman, *Stanley Park's Secret: The Forgotten Families of Whoi Whoi, Kanaka Ranch, and Brockton Point* (Madeira Park, BC: Harbour Publishing, 2005); Renisa Mawani, "Imperial Legacies (Post)Colonial Identities: Law, Space and the Making of Stanley Park, 1859–2001," *Law Text Culture* 7 (2003): 98–141; and Susan Roy, *These Mysterious People: Shaping History and Archaeology in a Northwest Coast Community* (Montreal: McGill-Queen's University Press, 2010).

23 For the history of environmental damage and restoration on the Duwamish River, see Mike Sato, *The Price of Taming a River: The*

Decline of Puget Sound's Duwamish/Green Waterway (Seattle: Mountaineers Books, 1997); and B. J. Cummings, *The River That Made Seattle: A Human and Natural History of the Duwamish* (Seattle: University of Washington Press, 2020).

24 Barman, "Erasing Indigenous Indigeneity," 9–10.

25 Lee Maracle, "Goodbye Snauq," in *First Wives Club: Coast Salish Style* (Penticton, BC: Theytus Books, 2010), 28.

26 Barman, *Stanley Park's Secret.*

27 Thrush, *Native Seattle*, 193–99; and Julia Anne Allain, "Duwamish History in Duwamish Voices: Weaving Our Family Stories since Colonization" (PhD diss., University of Victoria, 2014).

28 See *c̓əsnaʔəm: The City before the City*, dir. Elle-Máijá Tailfeathers (Vancouver: Moving Images Distribution, 2017).

29 A synopsis of *Ashes on the Water* can be found at https://ravenspiritdance.com, accessed January 21, 2020.

30 Visit the Squamish Nation website at www.squamish.net, accessed January 21, 2020.

31 Visit the Musqueam First Nation website at www.musqueam.bc.ca, accessed January 21, 2020. See also the Tsleil-Waututh First Nation website at https://twnation.ca.

32 See www.songheesnation.ca, accessed January 21, 2020.

33 See www.duwamishtribe.org, accessed January 21, 2020.

34 See, for example, the Muckleshoot Indian Tribe's website at www.muckleshoot.nsn.us, accessed January 21, 2020.

35 Daisy Couture, Sadie Couture, Selena Couture, and Matt Hern, *On This Patch of Grass: City Parks on Occupied Land* (Halifax and Winnipeg: Fernwood Publishing, 2018).

CHAPTER 6

A HISTORY OF PUYALLUP FISHING RESISTANCE

DANICA MILLER

ON WEDNESDAY, AUGUST 5, 1964, THE SENATE SUBCOMMITtee on Indian Affairs met in Washington, DC, to discuss Indigenous fishing rights and hear testimony from Indigenous fishers. Puyallup leadership was in attendance, including Vice Chairman Silas Cross Sr. and Chairman Frank Wright Sr. According to the introduction to the deposition, these hearings were "a most useful vehicle for initiating congressional study of the complex, and unfortunately sometimes inflamed, problem of the nature of Indian treaty rights to fish at usual and accustomed places."[1] The bureaucratic language used to describe fishing rights as "complex, and unfortunately sometimes inflamed" implies a disagreement between two parties and not, as history has shown, a state-sanctioned war against Indigenous fishing rights. In this chapter, I will look at the 150-year history of legal battles fought by the Puyallup to retain their ancestral and treaty fishing rights. This history is not exhaustive, as it would be almost impossible to collect every local, state, and federal record of arrest, court hearing, law, and policy for and against which the Puyallup have had to fight. Many of the laws involved several federally recognized tribes, and some laws were specific to the Puyallup. The persistence of the settler government in their determination to destroy the very spiritual and physical foundation of the Puyallup self was always met by the Puyallup people's continued collective resistance as the Puyallup politically adapted to the Western court system to assert their ancestral sovereignty (the right to self-determine on our land since time immemorial) within westernized constructs of nationhood (centralized government, written agreements, judicial system).

The following is an excerpted interview from 1964 between Frank Wright Sr., who served in Puyallup leadership capacities his entire life, and Senator Warren Magnuson, a lifelong Democrat who was considered sympathetic to Indigenous rights—a perception relative to how other settlers viewed Indigenous rights at the time.[2] A condescending tone pervaded the hearings, but the short deposition below exemplifies both the determination and grace of Puyallup fishers and their constant battle against settler law and settler mindset.[3]

FRANK WRIGHT SR. I know that they [sportsmen] are capable of taking any salmon.

SENATOR WARREN MAGNUSON. Oh, certainly on the Fraser River run up there.

WRIGHT. Well, even they fish all the way down in the sound. These sportsmen, they come around in the winter and they scowl at us and give us abusive names, and I read in this treaty that the Indians have the exclusive right on this land. These 600 fishermen and sportsmen that the director of game and fish spoke of were fishing on our exclusive fishing property rights. They are on the Indians' property and yet they are coming around and saying, "You leeches, you are taking our fish." We are not able to make things public in the papers, because we are just a minority and we would have to pay money to have statements made in the paper, so the whole minds of the public—

MAGNUSON. You have to what, pay money to have statements made in the paper?

WRIGHT. We would, yes. The minds of the public is swayed by the press so that the Indian is a leech—

MAGNUSON. I am sure that the *Tacoma News Tribune*, any time you have a statement, a reasonable statement, that they would be glad to publish it.

WRIGHT. My brother tried to make a statement in the paper and they refused it, so he had to send back to the American Society of Newspapers or something back east, and then he finally got it in our newspaper.

MAGNUSON. I am sure that would be an exception to the rule.

WRIGHT. Exception to the rule? You just try, it's quite a—

MAGNUSON. There is a reporter right here that will take your statement. [*laughter*] Now sometimes I have that trouble with newspapers, but that is because of the business I am in.

WRIGHT. I just made that statement because it shows that the Indian has very little chance of making himself known, and his side of the picture is not given.

MAGNUSON. Well, that is why we are all here. We are trying to find out as much as we can about this and to see where the equities lie and the facts and, as reasonable people, I am sure we can work it all out.

WRIGHT. We have asked for assistance in controlling our river like a statement was made that we do not back up our conservation measures wholly and fully, and we have met with this department of fisheries asking assistance. The only offer that we have had is they want complete control. They don't want to give assistance, they want control. Now, the Indian, as far as I can see, needs assistance but he knows what happens when control is given. So, about this meeting, one of the high points—

MAGNUSON. In other words, you say that the fisheries department wanted to give you assistance, but they wanted to handle their assistance inside. Is that right?

WRIGHT. Yes. "We will assist you if you let us take over."

In this telling deposition, Wright states the problems that the Puyallup encounter as fishermen. They have the treaty rights to fish, but sportfishermen not only violate these treaty rights but also verbally abuse the Puyallup. Wright continues to explain that the Puyallup have very little recourse to combat these violations because the media is not on their side either. Magnuson, who adds to Wright's argument with his constant interruptions, says that no one needs to pay to have their statements printed in the paper, which willfully disregards Wright's point. Wright's point is that the state fisheries and laws are on the side of where the money is, and, by extension, so is the media.

Wright continues to press his argument. The press is not on the side of the Indigenous, and in order for the press to be on the side of the Puyallup, the Puyallup would have to pay them, which rather slyly indicates that the settlers are paying the press, through advertisements and subscription. Magnuson, seemingly oblivious to the conversation in which he is participating, argues that it is perfectly easy to make a statement to the press as a Puyallup, which instead of supporting his own argument, just emphasizes his ignorance of media bias (and bias in general). He declares, "I am sure that the *Tacoma News Tribune*, any time you have a statement, a reasonable statement, that they would be glad to publish it," revealing his own ignorance about the history of the *Tacoma News Tribune* and how that particular newspaper took a decidedly anti-Puyallup stance from its earliest days of publishing until well into the 2000s, when the success of Puyallup gaming made it a politically unwise stance to take.

As the above deposition attests, the history of colonization of the Puyallup is a history of the Puyallup fighting for their ancestral and sovereign rights to fish in the face of the settlers' consistent eroding and erasure of Indigenous rights (see map I.2 in introduction). When I first began my research into Puyallup history, I was amazed not at the Puyallup's daily and dogged resistance (for I already knew that history from my family and my own experience) but at the persistence of the settler state. Why was the United States, and then later Washington State, determined to stop Indigenous fishing? Why were settlers threatened by a few Puyallups catching some salmon?

Having grown up with Puyallup fishermen my entire life, I knew they were not only the kindest, most quiet-humored people around, but they also had salmon. As I attempted to match the oral history with the Western documented one, I sat down with my father to clarify some of the issues confronting me. Finally, I asked him, "Why don't they want us to fish?" to which he replied, "Salmon is our everything. Without it, we are nothing. They know that, so they want to take it away from us."[4]

THE PUYALLUP AND EARLY SETTLER COLONIZATION

The period from 1846 to 1887, when the land that is now Washington State became Oregon Territory then later split off into Washington Territory in 1853, were years of aggressive violence toward

Indigenous people as settlers fought for control over Indigenous land. It is always important to note that during this territorial process, the United States claimed and settled land it had no right to. This land unequivocally belonged to the Indigenous nations. However, through supportive government policy, most notably the Oregon Donation Land Act of 1850, settlers arrived in the newly established territories—without treaties with Indigenous nations—ready to fight for land that they knew was not theirs but, by the history of settler colonialism as established by the United States, could be theirs if they were willing to violently battle for it.[5]

When settlers first started colonizing the region, they wrote of a paradise. But to the Puyallup, the land was not a gift from a white god but a sacred place that demanded respect and care; in return, the land provided all that was necessary. During initial settlement, despite settlers and Indigenous nations possessing vastly different concepts of land ownership and stewardship, the land continued to give in the ways it always had. Settlers, however, did not understand, nor did they take the time to understand, that the untouched paradise around them was the result of thoughtful land management by the Puyallup and interdependence based on love and respect.

The Puyallup understood that they were dependent upon the earth, and this dependence developed into a culture of love, respect, and stewardship toward the land. The land and water were managed by small clans with inherited rights to the fishing spots and gathering areas. This stewardship meant that areas were never overfished or overfarmed. In addition, the practice of traveling meant that goods that were difficult to grow could be obtained through trade with peoples whose land was able to support those resources.

The contradictory stance of the settlers is exemplified by the following quote from Governor Isaac Stevens in a letter to George Manypenny, then commissioner of Indian affairs:

> They [the Indians] catch most of our fish . . . supplying not only our people with clams and oysters but salmon to those who cure and export it. . . . Their principal food is fish and roots and berries. . . . The provisions as to reserves and as to taking fish . . . had strict reference to their conditions as above, to their actual wants, and to the part they play and ought to play in the labor and prosperity of the Territory. It may be here observed that their

mode of taking fish differs so essentially from that of the whites that it will not interfere with the latter.[6]

Stevens understood the process by which the Indigenous people catch fish and the importance of fish to the Indigenous people. He also comprehended, though perhaps not fully, how dependent upon the Indigenous people the settlers were for survival. The Indigenous people literally fed the settlers, not to mention supplied them with labor for industry and building of shelters. However, because the Indigenous peoples' method of catching salmon differed from the settlers' (Stevens implied the Indigenous method was less productive), their catch would not interfere with the fishing needs of the settlers. Stevens did not understand that he was witnessing Indigenous resource management and that, once the settlers began fishing in earnest, the salmon runs would begin depleting under the carelessness of capitalist industry.

Governor Stevens arrived in 1854 to quickly produce ten treaties with all the tribal nations of what is now Washington State. Stevens's career was one of impatient self-promotion.[7] The Puyallup (as well as the Nisqually, Squaxin, and others) signed the Treaty of Medicine Creek. Stevens did not take the Indigenous peoples' ancestral rights into consideration, which also means he did not consider or even recognize the stewardship they had always practiced in regard to the land. Stevens's team could not see that the land was cultivated; they could not see that trees were planted and pruned in ways that promoted their fruitfulness. Because the Puyallup managed the land with respect and without decimating it, the settlers did not understand that they were witnessing the literal fruits of Puyallup land management. The land was managed in order to sustain the Puyallup, and because the Puyallup understood their dependence upon the land, the land was managed in a way that benefited the land.

Governor Stevens understood very little about the people with whom he was signing a treaty and made little enough effort to do so. He did seem to understand that fishing was important to the Indigenous nations, and thus we have the oft-repeated Article 3 of the Treaty of Medicine Creek:

ARTICLE 3. The right of taking fish, at all usual and accustomed grounds and stations, is further secured to said Indians in common

> with all citizens of the Territory, and of erecting temporary houses for the purpose of curing, together with the privilege of hunting, gathering roots and berries, and pasturing their horses on open and unclaimed lands: Provided, however, that they shall not take shellfish from any beds staked or cultivated by citizens, and that they shall alter all stallions not intended for breeding horses, and shall keep up and confine the latter.[8]

However, access to fishing sites, as bounded by the stated reservation in Article 2 of the Treaty of Medicine Creek, was so terrible that war broke out immediately.[9] The original Puyallup reservation boundary made access to traditional fishing sites difficult, but for the Nisqually, the original reservation boundary made access to fishing sites impossible. The Treaty Wars, led by Chief Leschi of the Nisqually and many other warriors, foreshadowed the Fish Wars that would occur a hundred years later.

While it is outside the scope of this chapter to discuss the Treaty Wars, it is appropriate to note that the Treaty Wars ended with an executive order in 1857, which states in part:

> Governor Stevens has formed the opinion that the locations named in the first article of the treaty were not altogether suitable for the purpose of establishing Indian colonies. One objection was that they are not sufficiently extensive. . . . He has arranged a plan of colonization which involves the assignment of a much greater quantity of land to the Indians. . . . He proposes the enlargement of the Puyallup Reserve at the south end of Commencement Bay to accommodate 500 Indians, the change in the location, and the enlargement of the Nisqually Reserve.[10]

While this policy change rather conveniently recast Stevens as a benevolent servant of the Indigenous—and not as the cause of an avoidable two-year war due to his own arrogance and stupidity—it also marked the beginning of the Puyallup fighting for their fishing rights and winning, with the fight ending in federal policy change. The Puyallup used their ancestral sovereignty to assert their rights while fighting for them within a colonial system.

Washington became a state in 1887, which means that Washington was still a territory under federal jurisdiction in 1886 when federal troops were brought in to protect Indigenous fishing sites covered by treaties. Settlers had established large industrial fish canneries, which were beginning to deplete the salmon runs at the expense of Indigenous fishing needs. As a result, many Indigenous peoples accessed both their treaty-sanctioned fishing sites and traditional fishing sites outside of treaty boundaries in order to feed their families. In response, settlers began violently restricting Indigenous access to all fishing sites in order to facilitate greater runs for the industrial canneries. Before Washington was even a state, Indigenous people were blamed for the smaller runs of fish, and conservation was used as an excuse to deny Indigenous peoples their ancestral and Western sovereign rights.

THE PUYALLUP, WASHINGTON STATE, AND THE UNITED STATES

As a general rule, in the settler court system, tribes have usually lost on the state level and won on the federal level. States fight with the tribes over resources in which the federal government has little vested interest, and on the federal level, there is some pretense of honoring treaty rights as these are agreements made by the federal government itself. However, all treaties made with Indigenous tribes have been broken by the United States, so this pretense only goes so far. Often the state, in willful disregard of federal Indian law and treaty rights, follows its own agenda until forcibly made to adjust. Unfortunately, reaching the federal judiciary requires an amount of resources most tribes simply do not have. This general settler colonial history toward the Indigenous people of the now United States is certainly applicable to the Puyallup.

United States v. Winans in 1905 was an early win for Washington State tribes. *United States v. Winans* was filed on behalf of the Yakama and their rights both to access traditional fishing areas and to a guarantee that there would be fish to catch in those traditional fishing areas. While specific to the Yakama, this case had national implications and certainly was relevant and understood by all parties involved to apply to all Washington State treaty tribes. In writing for the majority, Justice Joseph McKenna stated that a "treaty between the United States and the Indians . . . is not a grant of rights to the Indians, but a grant of rights from them—a reservation of those granted."[11] This established for the first time the so-called "reserved rights" doctrine in federal Indian law. Treaty rights are not granted by the settler state but represent an agreement between the settler state and the Indigenous one. By outlining the "reserved rights" doctrine in *United States v. Winans*, the federal courts acknowledged this fundamental aspect of Indigenous treaty production for the first time—although Indigenous leaders and activists have always accepted this basic understanding.

The federal court ruled in favor of all Treaty of Medicine Creek tribes, that they had the right to access traditional treaty grounds. The court did add, however, that the state did have some rights in terms of regulation for the conservation of fishing. In response, under the rhetoric of conservation and in response to scarcity of fish, Washington State Supreme Court ruled in 1916 (twice) and in 1921 that Indigenous peoples had no greater fishing rights than whites and must comply with state law. Again, Indigenous peoples, who had practiced sustainability as a unique intertwining of spiritual and cultural beliefs, were blamed and punished for the careless and abusive exploitation of natural resources by settler capitalism. In none of these cases did the federal government honor its trust responsibilities and review this in federal court.

The Puyallup were well versed in their rights. From any sense of ancestral sovereignty, the Puyallup had the right to fish, but even within the limited viewpoint of the settlers, the Treaty of Medicine Creek clearly guaranteed the same. The Puyallup always fished. As salmon runs depleted, mainly due to industrial fish canneries, Puyallup fishing rights quickly became a target. With the rise of sportfishing in the 1940s, harassment and targeted violence toward Puyallup fishermen became the norm. In response, the Puyallup continued to do what they had always done: fish.

For the next three decades, state-sanctioned violence from sportfishermen and game wardens would increase toward treaty fishers. We commonly call this the Fish Wars. While I am focusing primarily on the Puyallup, the Puyallup Fish Wars were fought with, and were sometimes even indistinguishable from, the Nisqually battles. In addition, the Puyallup fishing activism that would erupt over the next decades was often part of Indigenous fishing activism involving many tribes, especially those tribes in Washington and Oregon. Concurrent to this, the rise of the civil rights movement influenced Indigenous activism, which saw a meteoric rise in attention with various media events, most notably with the takeover of Alcatraz and American Indian Movement initiatives.[12] All of these larger and various movements guided the Puyallup, though many Puyallup Elders today specifically acknowledge local chapters of the Black Panthers for their sustained, invaluable assistance with helping the fish warriors organize against state violence. Yet although it is important to contextualize the Fish Wars within a larger historical moment, it is also important to emphasize that in many ways, Puyallup fisherman had been a part of these battles since settler colonization began. To fight for their right to fish became inherent to the tradition of fishing, no matter the national and state political climate.

On a local level, fishers often had their boats filled with cement by sportfishermen. Arrest could also mean confiscation of boats and equipment, which could easily represent a year's salary for Puyallup people at that time. There are reports from early in the Fish Wars of dugout canoes confiscated by the state. A dugout canoe is made from a hollowed tree trunk, which is the traditional and spiritual—not to mention labor-intensive—way of canoe building. In the face of increasing violence and policing, treaty fishers continued to fish, mostly out of the continued need to support themselves through traditional means but sometimes in order to provoke arrest in hopes of getting local laws changed or federal laws upheld.

In 1954, Robert Satiacum and James Young of the Puyallup were arrested and charged with possession of game fish during a closed season and with use of fixed gill nets to catch game fish at two locations. According to court proceedings, law enforcement officers observed Young tending two fixed nets on the Puyallup River within Tacoma city limits. He was not arrested, but the officer observed he had two salmon with him. The next day, the same officers saw Satiacum and Young again tending the nets. They had three steelhead and were promptly arrested. One location was outside the Puyallup reservation, and the second was within the original boundaries of the reservation but on land now owned by settlers. Of course, both locations were on ancestral Puyallup land that still belongs to us, regardless of settler laws and treaties. Both men knew these acts were illegal in the state, and both also knew they were acting within their Puyallup rights to fish. They were committing acts of civil disobedience, but they were also feeding their families. Washington Superior Court ruled in the state's favor.

The Puyallup appealed to the Washington Supreme Court, which supported the earlier ruling only if the state could prove that the regulations were in place for conservation efforts. At this time in Washington State, there were about 800 treaty fishers, 6,600 non-Indigenous commercial fishermen, and 283,000 sportfishermen. In addition, 4,000 oceangoing trawlers worked in the waters, each year hauling in about two million tons of salmon.[13] The state was hard-pressed to prove that Puyallup fishing was the cause of salmon depletion. The case was dismissed "for want of sufficient evidence as to the reasonableness and necessity of the statutes and regulations involved for the conservation of fish." While this was an important win for the Puyallup, Washington State still maintained authority over the Indigenous fishers, and so Indigenous fishers continued to fight for their treaty rights.

In his testimony to the Senate Subcommittee on Indian Affairs in 1964, Frank Wright Sr. rightly complained about the lack of media attention toward Indigenous fishing rights. In response, Indigenous fishing activists were beginning to use the Western media to their advantage by staging "fish-ins." The tribes' relationship with actor Marlon Brando, an outspoken ally of Indigenous rights, is a prime example of how the "fish warriors" began to use the media to garner attention toward the annihilation of their rights. On March 2, 1964, Brando and Satiacum went fishing along with invited media in order to provoke arrest. In photographs, the two men mock Western policing of Indigenous fishing by hamming it up for the newspaper, smiling and waving at the cameras. In an image from the *Seattle Post-Intelligencer* archives, we see them baiting the Department of Game and Wildlife to arrest them for violating state gaming laws (even though those same game laws violated the federal Treaty of Medicine Creek).[14] Satiacum and Brando were clearly enjoying themselves, and for all intents and purposes, so was the media. Here was one of the biggest stars of the day fraternizing with some Indian toughs, and all of them were about to be arrested, and for what—fishing? This is what Washington State was spending millions of dollars to stop? And arrested they were, for violating fishing laws, but the joke was on the state—and to some extent, the media—because the men were not actually fishing but performing: Brando's fish was frozen. Their fishing performance drew much-needed media attention to the Puyallup and Nisqually and garnered national public sympathy for their cause, something that Washington State definitely did not want.

The City of Tacoma spent the 1960s and 1970s aggressively targeting the Puyallup and dismissing their continued demands to practice their sovereign rights. In May 1965, Judge John D. Cochran of the Pierce County Superior Court declared the Puyallup Tribe no longer existed because the reservation no longer existed. This was despite the fact that two pieces of trust land were still held by the Puyallup and several pieces of land, which had fallen out of trust but were still owned by the Puyallup, also existed.[15] Additionally, the court ruled that treaty rights were not absolute, and those treaty rights certainly did not extend "to take fish with such gear and at such times as would destroy the fishery."[16] Understandably, the Puyallup Tribe had issue with this ruling. Not only were their treaty rights nullified, but their very existence was as well. Washington State officials were clearly aware that a state did not have the right to remove federal recognition from a federally recognized tribe, but they seemed to be betting on both Puyallup ignorance and Puyallup powerlessness. Much to the relief of the Puyallup, who thought they existed but were told otherwise by Pierce County, the Washington Supreme Court asserted in 1967 that the Puyallup Tribe did in fact exist and that such an injunction against the tribe was discriminatory. However, the existence of the Puyallup reservation remained undetermined, as did the question of whether set nets could be used. Consequently, the state continued to close the Puyallup River to Puyallup fishermen, who still continued to fish the river and thus continued to be violently harassed for practicing their sovereign rights.

On May 27, 1968, the Puyallup Tribe took Washington State to the Washington Supreme Court in *Puyallup Tribe v. Department of Game*, arguing for their right under the Treaty of Medicine Creek "of taking fish at all usual and accustomed grounds and stations." The court determined that the state could regulate the tribe's fishing rights, including the traditional use of set nets for fishing steelhead and salmon, and sent the case back to trial court to determine if the conservation regulations were reasonable and necessary. The US Supreme Court, splitting the difference, determined that in the interest of conservation, the state could regulate Puyallup fishing "in common with" fishing by others—meaning, equitable with the settler allocation. The state was told to determine a formula for allocation of steelhead between the Puyallup and sportfishermen that would be fair to both. This equity was problematic as it did not necessarily mean "equal" but rather whatever the state would determine to be fair.

In the summer of 1970, the fish warriors faced their biggest battle. A small piece of the Puyallup riverbed was found to never have been stolen by settlers and to still remain in Puyallup ownership. The Fish Wars had been largely fought in the Nisqually and Puyallup Rivers, but to show unity across the tribes as well as to maximize media exposure by focusing on a large urban area, the warriors moved to a densely populated area along the Puyallup River and set up an encampment. The area along the Puyallup River was one of the few unceded lands of the Puyallup and had never been allotted out to an individual Puyallup. The only other communally owned Puyallup land at the time was the Puyallup cemetery. With the fish-in, the tribes were asserting their ancestral sovereignty on land that was tribal land—by all accounts. The fish warriors spent the summer of 1970 fishing, laughing, and exchanging taunts with heavily armed local police and game wardens, who harassed the encampment with their intimidating, militarized, and state-sanctioned threat of violence. Then, on September 9, 1970, the morning erupted in battle. Fighting broke out between the warriors and the local police and game wardens. The smoke could be seen for miles. Over sixty local Indigenous activists were arrested by extremely violent methods. An Indigenous child as young as eight was taken into custody. Nine days after the battle, the Justice Department filed suit against the State of Washington on behalf of the tribes in *United States v. Washington*. The fishing question would be settled hopefully once and for all.

In *United States v. Washington*, northwest tribal communities argued that according to the Stevens treaties, they had the right to fish without state restriction.[17] Much to the astonishment of almost everyone involved, the very conservative Judge George Hugo Boldt agreed, in a judgment now commonly known as the Boldt Decision. Boldt referenced the language of the treaties, especially the previously quoted Article 3 of the Treaty of Medicine Creek. According to Boldt's research team, nineteenth-century legal dictionaries defined "in common with" to mean "equal resource distribution," which he interpreted to have been the intent of the Treaty of Medicine Creek at the time of signing. Judge Boldt ruled that according to the treaties, local tribes had the right to 50 percent of the fish harvest. He also declared that all state regulation of Indian fishery that did not apply to conservation was illegal. And finally, he determined that only Congress had the authority to derecognize tribes or tribal reservations, and thus the Puyallup reservation still existed, despite the best efforts of the City of Tacoma. The controversial ruling (for non-Indigenous state and local governments, that is) took several years to implement, especially regarding shellfish harvests. Whether tribes had the right to shellfish harvests and in what capacity became another lengthy battle that would remain unresolved for another twenty years, until the federal government ultimately ruled in the tribes' favor again.[18]

The Puyallup fight for their fishing rights highlights the dichotomies inherent in constructing tribal sovereignty today. The Puyallup's fight in the courts for their ancestral sovereignty acknowledges both this sovereignty and, at times, the tribe's dependency upon the settler government. And yet there are multiple sovereignties (as defined by Western and Indigenous systems), and the Puyallup continue to construct, manipulate, and fight for their sovereignty by using this multitude of frameworks.

CONCLUSION

We fish because it is who we are, and we fight to continue to be who we are. We practice our Indigenous sovereignty through fishing. Fishing is of us, borne of our ancient belief system; we have the innate right to support ourselves through this practice. As Frank Wright Sr. testified in the earlier-referenced hearing:

> Since salmon was 80 to 90 percent of their diet, the Puyallup held a cultural festival or religious ceremony in honor of the salmon. At this ceremony, they barbecued the first salmon of the spring over an open fire. It was then parceled out to all, in small morsels or portions so all could participate. Do this, all bones were intact. Then, in a torch-bearing, dancing, chanting, and singing procession, they proceeded to the river, where they cast the skeleton of the salmon into the stream with its head pointing upstream, symbolic of a spawning salmon, so the run of salmon would return a thousandfold.[19]

There are many important pieces of Puyallup life in this short passage. Within a Puyallup lens—a lens of practical spirituality wedded to generous welcome—the blur between religion and culture is embedded in salmon, which sustains us physically, spiritually, and culturally, between which there can be no real distinction. We shared the salmon with all so we could all participate, because we understood that even though we were as generous as the land was, we were also comfortably reliant on one another. Then in a spiritual act both serious and joyful, the Puyallup proceeded to the river where they returned the salmon to its ancestors as a sign of respect. The head pointed upward, like all spawning salmon, symbolized more salmon to return. We knew, either consciously or through immemorial habit, that returning the bones to the river also ensured the health of the river through this compostable practice.

To this day, the Puyallup continue to practice the First Fish Ceremony, and they continue to return all salmon bones to the river. We take care of the river, and the river takes care of us.

NOTES

1 *Indian Fishing Rights: Hearings on S. J. Res. 170 and S. J. Res. 171, Before the Subcommittee on Indian Affairs of the Committee on Interior and Insular Affairs*, 88th Cong. (1964).

2 When Magnuson lost his senate seat to Republican Slade Gordon, who ran his campaign on an anti-Indigenous fishing rights platform, it was largely viewed as a public rebuke of the recognition of Indigenous treaty rights.

3 *Indian Fishing Rights* (testimony of Frank Wright, member of the tribal council, Puyallup Tribe). The transcript has been edited for clarity.

4 Bill Sterud, Vice Chairman of the Puyallup Tribe of Indians, personal conversation with the author, 2015.

5 Alexandra Harmon, *Indians in the Making: Ethnic Relations and Indian Identities around Puget Sound* (Berkeley and Los Angeles: University of California Press, 1988), 58.

6 Vine Deloria Jr., *Indians of the Pacific Northwest: From the Coming of the White Man to the Present Day* (1977; Golden, CO: Fulcrum Publishing, 2012), 43.

7 Richard Kluger, *The Bitter Waters of Medicine Creek: Tragic Clash between White and Native America* (New York: Knopf, 2011), 67.

8 Treaty with the Nisquallys, etc, Dec. 26, 1854, 10 T.S. 1132.

9 Here is the text of Article 2 of the Treaty of Medicine Creek:

> There is, however, reserved for the present use and occupation of the said tribes and bands, the following tracts of land, viz: The small island called Klah-che-min, situated opposite the mouths of

Hammersley's and Totten's Inlets, and separated from Hartstene Island by Peale's Passage, containing about two sections of land by estimation; a square tract containing two sections, or twelve hundred and eighty acres, on Puget's Sound, near the mouth of the She-nah-nam Creek, one mile west of the meridian line of the United States land survey, and a square tract containing two sections, or twelve hundred and eighty acres, lying on the south side of Commencement Bay; all which tracts shall be set apart, and, so far as necessary, surveyed and marked out for their exclusive use; nor shall any white man be permitted to reside upon the same without permission of the tribe and the superintendent or agent. And the said tribes and bands agree to remove to and settle upon the same within one year after the ratification of this treaty, or sooner if the means are furnished them. In the meantime, it shall be lawful for them to reside upon any ground not in the actual claim and occupation of citizens of the United States, and upon any ground claimed or occupied, if with the permission of the owner or claimant. If necessary for the public convenience, roads may be run through their reserves, and, on the other hand, the right of way with free access from the same to the nearest public highway is secured to them. (Treaty with the Nisquallys, etc)

10 In a letter to President Franklin Pierce, Commissioner George Manypenny shared a report from Governor Isaac Stevens, in which he requested enlarging the size of the Puyallup and Nisqually reservations. President Franklin Pierce responded in the positive. "Puyallup Reserve," in *Executive Orders Relating to Indian Reserves* (Washington, DC: Government Printing Office, 1902), 131–32.

11 United States v. Winans, 198 U.S. 371 (1905).

12 Paul Chaat Smith and Robert Allen Warrior, *Like a Hurricane: The Indian Movement from Alcatraz to Wounded Knee* (New York: New Press, 1997).

13 Alvin M. Josephy Jr., *Now That the Buffalo's Gone: A Study of American Indians* (Norman: University of Oklahoma Press, 1984), 189.

14 Photograph of Marlon Brando and Bob Satiacum on Puyallup River, March 2, 1964, Seattle Post-Intelligencer Photograph Collection, available at David Wilma, "Native Americans and Supporters Stage Fish-In to Protest Denial of Treaty Rights on March 2, 1964," HistoryLink.org, March 1, 2003, www.historylink.org/File/5332.

15 City of Tacoma, Wash. v. Andrus, 457 F. Supp. 342 (D.D.C. 1978).

16 Josephy, *Now That the Buffalo's Gone*, 196.

17 United States v. State of Washington, 384 F. Supp. 312 (W.D. Wash. 1974), aff'd, 520 F.2d 676 (9th Cir. 1975).

18 United States v. State of Washington, 898 F. Supp. 1453 (W.D. Wash. 1995).

19 *Indian Fishing Rights.*

CHAPTER 7

OUR RIVER, OUR FUTURE

More-Than-Local Grassroots Activism in the Portland Harbor

ERIN GOODLING

THE OPPORTUNITY TO PURSUE URBAN SWIMMING AND KAYAKING in the Willamette River is a key feature of the city of Portland, Oregon's recent sustainability success story—and to its celebrated sense of place. For the past decade or so, "riverlution" boosters have deemed the city's main urban waterway a perfect place to cool off on hot summer afternoons. In spring 2016, Portland mayoral candidate (and soon to be mayor) Ted Wheeler even delivered his ballot via "tiny triathlon." He ceremonially swam across the Willamette, biked along the waterfront, walked a few blocks to Pioneer Courthouse Square, and dropped his voting card in the box.[1] Mayor Wheeler's display of outdoor recreational aptitude no doubt appealed to legions of voters with similar passions for the city's place-based amenities, including the harbor.

Portland has been called the bike capital of the United States for its miles of dedicated bike lanes. Lonely Planet listed the city as one of its top twenty destinations of 2017 on account of the city's eco-friendly activities, along with nearby world-class skiing, hiking, and surfing.[2] Part and parcel of Portland's reputation as an outdoor recreation mecca, the city is renowned for its leadership on environmental matters. Cities around the world, for instance, seek to emulate Portland's award-winning climate action plan.

Yet, despite Portland's reputation as a city with a sustainability-oriented sense of place that benefits all residents, the reality is less romantic. Following his first mini-triathlon, after taking office, Mayor Wheeler took another public dip. He remarked, "Today we're going to swim in the water, the water quality is very good. We're not going to stop and eat mud on the bottom of the river."[3] Wheeler's statement sums up conventional narratives of the Portland Harbor. Sustainability boosters, the media, planners, and policymakers by and large offer a storyline about the harbor that goes something like this: *For much of the twentieth century, the Portland Harbor was a cesspool of refuse from municipal sewage and stormwater runoff and from upriver pulp and paper industry pollution. Following tighter regulations and completion of massive public infrastructure investments, including the $1.4 billion "Big Pipe" in 2011, toxic runoff into the Willamette River has dramatically declined. The river is fine!* For the majority of Portland residents, the toxins in the mud pose little immediate danger, as Mayor Wheeler suggests. The river is indeed clean enough for kayaking and swimming, most of the time, for most people. But industrial toxins buried in a ten-mile stretch of the river's soil and sediment actually prevent thousands of people from safely engaging in cultural traditions and life-sustaining activities. In 2000, the Environmental Protection Agency (EPA) listed the harbor as a Superfund site, prioritizing federal oversight of its cleanup (see map 7.1).

Since 2012, the Portland Harbor Community Coalition (PHCC) has been fighting for a more thorough—and just—cleanup of the Portland Harbor. PHCC connects people from a wide variety of races, ethnicities, and class backgrounds and with a range of motivations for seeking environmental change. As a central part of PHCC's activism, participants have produced a new, shared narrative about the harbor's past, present, and future. In June 2016, a few months prior to the mayor's declaration of the river's health, Rahsaan Muhammad spoke for PHCC at a press conference, condemning the EPA's watered-down remediation plan:

> We are here today representing over a dozen organizations . . . that are saying NO! We want a comprehensive cleanup. . . . What they are proposing is truly insufficient. . . . There are billions of dollars that have already been pocketed by corporations like Bayer CropScience, Chevron, Gunderson, Exxon, BP, Shell, Schnitzer Steel, Toyota, the US Navy, just to name a few—and we all can

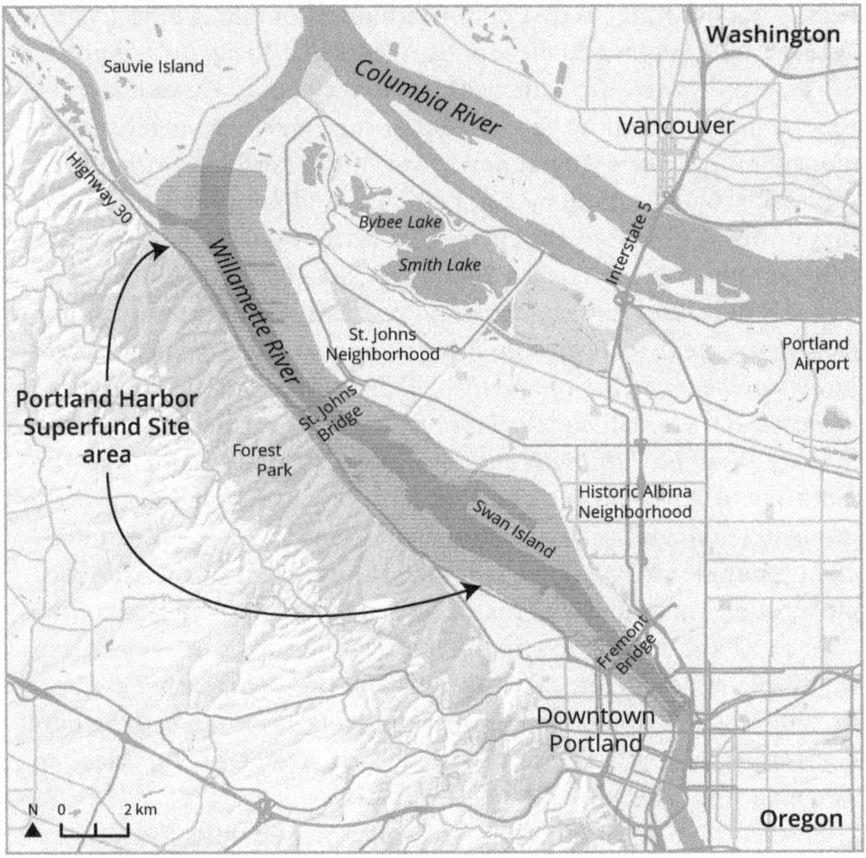

MAP 7.1. The Portland Harbor Superfund site and surrounding area. Map by Dillon Mahmoudi.

> look at their records and what they make yearly and see that that 750 [million dollars] that you all are playin' with, tryin' to clean up the harbor, is really an insult and a joke. We are here, men, women, and children, ALL different nations—make no mistake about this—this coalition is representative of ALL the communities of color, ALL the Native tribes. . . . We are ALL on board with *cleaning up* our harbor *at the expense of those* who are responsible for *polluting* the harbor.

By asserting that the waterway is far from healthy, notwithstanding infrastructure improvements that have enabled the river to become an epicenter of urban recreation, Rahsaan's speech conveys a much more complex narrative of the harbor and its history than do mainstream storylines.

Grassroots groups led by Black and Indigenous people, immigrants and refugees, and houseless people of diverse backgrounds make up PHCC's core membership. Coalition supporters include environmental, labor, and neighborhood groups. One might expect that those pushing hardest for a cleaned-up harbor would be those living closest to the river. After all, "place attachment," which entails an affective bond—an emotional connection—between people and place, often precedes work to preserve or improve a particular place. People typically come to intimately know a place by spending time there, deepening their personal and ancestral ties over years, decades, centuries. But, with the exception of houseless people living along the river, few PHCC members actually reside near the Portland Harbor today. Most hail from throughout the city and surrounding areas, and a few even live hundreds of miles away from the metropolitan center. Many PHCC members travel to the harbor for fishing, water ceremonies, picnics, cookouts, music festivals, canoeing, and kayaking. Yet some rarely set foot in the harbor, and others invoke rivers of their youth in places far away as they participate in PHCC's functions. The sociospatial complexity of PHCC's membership and resistance efforts is fundamental to grasp in order to understand who is shaping the Portland Harbor, and the Cascade region more broadly, from below.

This chapter addresses how Portland became famous for sustainability while refusing to fully attend to pollution in the river running right through the heart of the city—and how people have joined together to push back. As do sites in other Pacific Northwest cities, the case of the Portland Harbor brings into focus who benefits from, and who pays for, urban sustainability transitions. It also poses key questions about the future of green cities: Who is leading resistance efforts toward a more *just* sustainability?[4] How do the "geographies of resistance" that those on the front lines represent complicate understandings of place? Examining PHCC's efforts also helps illuminate the complex ways in which place attachment develops and drives collective work for a more just, green future. For some people, spending time in a given place instills a sense of connection and in turn motivates responsibility to work for the future of that place. But, for many, place attachment develops despite physical separation from or forced migration to a place. Although they may be "taking place in space," asserts Black studies scholar Katherine McKittrick, land relations are complicated by "subjectivities, imaginations, and stories."[5] Symbolic relationships, then, rooted less in a physical connection but rather in intangible connections to land, water, and other landscape features, can contribute to shared meanings of a place—which in turn propel collective efforts to defend and shape the future of a place.[6]

While PHCC's efforts are resolutely place based, it is not only a time-oriented, physical connection but also a symbolic, relational, more-than-local place-based-ness, with PHCC members' work today coming into conversation with peoples and rivers of other times and places. If public agencies are to adequately respond to frontline communities, they must take this relational place attachment seriously. In the meantime, groups such as PHCC and others throughout Cascadia will continue to push their own narratives in fights for a greener, and more just, future.

To understand PHCC's more-than-local, relational understanding of place, I draw on five years working as an organizer and activist-researcher with PHCC, from 2013 to 2017, as well as interviews with PHCC members conducted in 2015. These experiences are informed by my own white settler family's ties to the harbor. I grew up hearing my Irish Catholic great-grandfather's tales of fishing on the Willamette River as a boy and my uncle's stories of working in the shipyards. I also spent summers helping my Hungarian immigrant family tend our vegetable garden on the banks of the Columbia Slough, a Willamette tributary. Though I was no stranger to Portland's waterways when I got involved with PHCC, I had much to learn from those who have shaped and been shaped by the harbor for generations. One of my roles has been to help stitch together different historical threads into a cohesive narrative of the Portland Harbor, from a "peoples' view": from the perspectives of those who have carried the burdens of pollution, exploitation, dispossession, and displacement for generations—and who have long worked for a new way of surviving and thriving.[7]

In this chapter, I first share a brief, collectively produced history of the Portland Harbor from the vantage of frontline communities. This history provides a foundation to then articulate the recent origins of PHCC and its members' motivations for joining the coalition—which largely do not depend upon living in close proximity to the harbor today. I then contrast PHCC members' and EPA officials' conceptualizations of "place." In 2014, it became clear that sharply diverging ways of thinking about place and place attachment, and in turn who has been most impacted by the harbor's contamination, would have to be reconciled if the EPA was to respond to the needs of PHCC's members. The chapter ends with a brief discussion of how public agencies might adjust their stance to better respond to frontline communities.

A PEOPLES' HISTORY OF THE PORTLAND HARBOR

To understand why PHCC members and supporters are not celebrating the Portland Harbor as an urban recreational paradise and to grasp why people from such diverse geographies fight for improvements to a place in which they may never live, it is necessary to examine the history of the harbor from the perspectives of those who are fighting for its future. The very brief history presented here is one version of a longer collective narrative of the harbor's past that PHCC members and supporters have stitched together over the years, through sharing personal memories at monthly member meetings; attending one another's culturally specific events and ceremonies along the river; and developing op-eds, short films, a comic book, and more.

Centrally, PHCC's narrative emphasizes the endurance of Indigenous peoples in the Willamette Valley, despite exposure to disease, massacre, forced removals, and the destruction of traditional food habitats. "We've always been here," explains JR Lilly (Diné/Navajo), a PHCC member and former representative of the Portland Youth and Elders Council.[8] Prior to the 1800s, approximately three thousand Chinookan people belonging to multiple bands and tribes called the present-day Portland metropolitan area home. An abundance of resources, especially wapato, a potato-like root vegetable, as well as Chinook salmon, camas bulb, a variety of berries, and much more, made Wapato Island (today's Sauvie Island) and the rest of the valley suitable for year-round residence. The Clackamas band of the Chinook lived between Willamette Falls and the Willamette's confluence with the Columbia River. The Multnomah lived mainly in sixteen villages on Wapato Island. The valley's food sources also attracted many more tribes from around the region for seasonal trading, fishing, and gathering, including the Kathlamet, Tualatin Kalapuya (Atfalati), Molala, and Klickitat. Dozens of Chinookan villages, including those of the Multnomah as well as the Kathlamet, Clatsop, Kalama, and others, dotted the length of both banks of the Columbia River from present-day Bonneville Dam to the Pacific, and a few were located to the north and south of the Columbia's confluence with the Pacific, on the coast; these groups traveled periodically to today's Portland basin. Altogether, there were at least fifty-five documented Chinookan villages along the region's waterways, some inhabited year-round and many more seasonally.[9] Governance of water and land throughout the Willamette Valley and beyond was largely according to usufruct terms: that is, various people depended upon a site or resource without owning it. Whole communities, whether neighboring or through kinship ties, had usufruct rights to fishing sites, camas fields, wapato-root ponds, shellfish beds, and other sources of food and shelter.[10]

With the arrival of white settlers in the late 1700s, Chinookans' way of life was severely disrupted. Colonial explorers seeking otter pelts for Britain and the United States to trade with China brought diseases, including smallpox and malaria, reducing the Willamette Valley Indigenous population from fifteen thousand to fewer than two thousand people. Denying the humanity of Indigenous people, white settlers also murdered those who stood in the way of land and resources. And beginning in the 1850s, the US government ratified treaties that eliminated access to traditional homelands, displacing thousands to reservations around the region. Many hid for years, and others fought back directly, such as in the Rogue River Wars of 1855 to 1856. Nevertheless, such violent acts opened space for white Euro-Americans to establish the Portland Harbor as a main hub for the exportation of grain, lumber, and other commodities and, later, as a primary site of wartime shipbuilding and shipbreaking. Chinookans and many other Indigenous peoples continue to call the region home and are working to heal from over two centuries of colonial genocide.[11]

In addition to recognizing the past and present of Indigenous people, PHCC's narrative centers the persistence of Black shipyard workers and their families, notwithstanding workplace exploitation, exposure to industrial pollution, segregation, discrimination, and serial displacement over the last several decades. Whereas white people flocked to the state under the Oregon Donation Land Act (ODLA) of 1850, a clause in the state's constitution excluded people of African descent from crossing state lines, under threat of the lash. The ODLA "promised free land to white settlers only," codifying efforts to create a "white homeland" up until 1926. A handful of Black people migrated to Portland in the late 1800s and early 1900s despite these discriminatory laws.[12]

It was not until World War II that large numbers of Black people came to Portland, to work in the shipyards. Portland would become home to the largest shipbuilding and shipbreaking operation in the United States, thanks in part to the labor of more than twenty thousand Black workers, many thousand Indigenous people from across the country, and many Chinese migrants. Over the years, workers were exposed to extremely toxic substances on the docks, such as lead and asbestos. Black households were also exposed to diesel emissions from harbor and freeway traffic in the nearby segregated Albina district. For the first few years following mass migration to Portland, many Black Portlanders lived in Vanport, a city built for shipyard workers at the north end of the harbor. Following a 1948 flood, however, Black families were left with few options but Albina due to bank redlining, racial covenants in white neighborhoods, and discriminatory landlords and real estate agents. Compounding this violence, Black shipyard workers were not allowed to join most shipyard unions, which meant lower wages and fewer workplace protections than for their white counterparts.[13]

Starting in the 1920s, just as the shipbuilding and breaking industry was gaining steam, public health experts, sanitary engineers, conservationists, and well-to-do anglers began a fight to clean up the Willamette. These early environmentalists' concerns revolved less around shipyard contamination, however, and more around municipal sewage and toxic discharges from upriver pulp and paper mills. Advocates worried about the harmful effects of such toxins on commercial fishing as well as on tourist and recreation-centered fishing, business, and swimming. In 1938, they succeeded in establishing the State Sanitary Authority, which mandated that pulp and paper industries stop dumping waste into the Willamette. Then, between 1947 and 1952, state-level officials impelled the City of Portland to spend $15 million (the equivalent of nearly $147 million in 2021) on a wastewater treatment plant.[14] The plant was sited at the north end of Albina, Portland's segregated Black neighborhood.[15]

These remediation efforts helped put Portland on the map as a leader in the nascent environmental movement, and in 1972, *National Geographic* even featured stewardship of the Willamette in a cover article. Later, the city's $1.4 billion investment in sewer and stormwater infrastructure—the Big Pipe—substantially reduced sewage overflow into the Portland Harbor. Water quality in the harbor has improved substantially, and swimming is now considered safe most of the time. But Portland's sustainability successes, including river remediation and green infrastructure improvements, have benefited middle- and upper-class—largely white—residents far more than other groups; as publicly subsidized green-themed development has driven up real estate prices in inner Portland neighborhoods, including Albina, Black Portlanders in particular have been pushed out.[16] Building on decades of Black-led community organizing, numerous grassroots groups, including PHCC, are now pushing back against displacement.[17]

In conversation with this Indigenous and Black-centered history and present, PHCC also recognizes the ongoing exposure of food-insecure subsistence fishers, including immigrants and refugees, to toxins (see figure 7.1). Joining Indigenous and Black fishers, people from Eastern Europe, Southeast Asia, and Central America fish in the harbor on a regular basis. Many are continuing cultural traditions begun in their homelands and relish teaching their children to fish. In the midst of today's extreme housing unaffordability, fishing also provides some relief against food insecurity. Some people realize the potential hazards and continue to fish for subsistence and cultural reasons, whereas others are unaware of the risks.

Finally, PHCC understands that houseless people of diverse races and ethnicities seek refuge along the river, as they have periodically for over a century in Portland's main watershed. Their lives are threatened when they consume contaminated fish, live on toxic soil, and experience police violence. An estimated thirty-eight thousand people experienced houselessness in the tri-county area in 2017, with Black and Indigenous people disproportionately represented.[18] As police evict more and more houseless people from inner-Portland neighborhoods, the river maintains its draw for those seeking

FIGURE 7.1. Signs throughout the Portland Harbor warn against consuming resident fish. Photo by author.

stability. PHCC members worry, however, that riverbank "sweeps" will intensify in the years to come as federally mandated Superfund cleanup progresses and gentrifying (re)development plans unfold, at the same time as ongoing exposure to toxic soil and fish continues to negatively impact houseless people due to cleanup delays.

This brief "peoples' history," focused primarily on Indigenous, Black, immigrant and refugee, and houseless communities' experiences, amends conventional narratives of the river's cleanup. Make no mistake: PHCC recognizes that past remediation has laid an important foundation for continued decontamination. But like mainstream national narratives, which are lacking not in their "facts, dates, or details" but rather in their "essence," to borrow the words of Indigenous scholar Roxanne Dunbar-Ortiz, conventional histories of the Portland Harbor obscure who has carried the burdens of harbor pollution.[19] Conventional histories also obscure who has suffered severely as a result of Portland's greening thus far and who may be further excluded from the city without explicit anti-displacement provisions. Such whitewashing sidesteps questions of reparations and restitution. Counternarratives, like those produced by PHCC, begin to do the opposite, serving to affirm the existence and experiences of thousands of people with complex connections to the harbor.

PHCC'S ORIGINS

Members of the communities highlighted in this peoples' history—Indigenous people, Black people, immigrants, refugees, and diverse houseless people—came together to form PHCC in 2012. The origins of the coalition stretch back more than a decade prior, though. In the early 2000s, federal officials invited a local group focused on air quality, the Environmental Justice Action Group (EJAG), to join the EPA's formally sanctioned Portland Harbor Community Action Group (CAG) and take a central role in the cleanup planning process. At first, EJAG leaders appreciated the opportunity to participate. But EJAG did not stay involved for long. "We went to help them—but they didn't really listen," remembers Jeri Jimenez, a Klamath tribal member and EJAG's director at the time. To Jeri, the EPA's convening felt more like an official check mark on a public participation worksheet than a process that would truly allow for community members to inform the Superfund decision-making process. EJAG leaders withdrew from participation.

In the meantime, Superfund cleanup planning moved forward with little input from frontline communities. Starting in late 2011, polluting companies began doing their own outreach. At this point, Jeri became even more concerned that members of vulnerable groups were being excluded from the cleanup planning process. A phone call in winter 2012 from a representative of a local Latinx-serving nonprofit left Jeri troubled that impacted communities were being included in order to be paid off, to adopt the perspectives of the polluters. The caller described outreach that the Portland Harbor Partnership (PHP) had conducted through a public relations firm. The PHP was made up of a group of nine "potentially responsible parties" (PRPs), companies potentially on the hook to pay for cleanup. Jeri recalls the conversation: "He said, 'I think they are trying to bamboozle us.'"

Jeri decided to attend some of the outreach events. In addition to severely misleading audience members, she remembers that one presentation—to mostly Spanish speakers—was entirely in English and that it obscured who would be implicated in paying for the cleanup. "[The presenter] dumbed down his presentation, with the end of it saying, 'Well, you're the taxpayers. It's going to fall on you, so whether we spend a million or a hundred million, it's going to fall on you, so it's your decision.' And I'm like, 'Oh my God!'" Jeri understood that local resident and business ratepayers might be required to pay some of the cleanup costs, given that the City of Portland was a PRP, but she also knew that the full cleanup tab would likely be shared among dozens of mostly multinational corporations. Jeri emphasizes, "It was a really stupid presentation. It was *so* dumbed down." For Jeri, the presentation was nothing but a scare tactic to convince residents to advocate for a less costly cleanup.

The last straw came when one of the PHP's presenters advised a group of people who consumed resident fish from the Portland Harbor on a regular basis to essentially imagine away any possibility of bodily harm. "Don't think of them as chemicals," Jeri recalls the PHP's representative suggesting of the deadly toxins buried in the river's sediment. He suggested the audience instead think of these pollutants as "little bugs in the mud." Jeri could not believe what she was hearing: "My biggest concern was, they're out there, flimflamming communities of color, and [the PRPs] spent a whole bunch of money to create a lie that [they] think will work."

Jeri immediately called Cassie Cohen, a white woman who was the director of Groundwork Portland (GWPDX), a fledgling local chapter of an organization focused on youth and environmental justice (EJ). Jeri felt that with a home base in Albina, Portland's historic Black neighborhood, and given that Cassie had developed relationships in years past with several grassroots groups, GWPDX was uniquely positioned to convene community groups around the Superfund cleanup. GWPDX's board was majority Black, and unlike most other social and EJ-focused organizations in Portland, GWPDX's work concentrated on community-controlled remediation and reuse of brownfield sites; the board supported outreach to other frontline communities.

Cassie made the connection to the Superfund cleanup: "[Groundwork] did brownfields work, and Superfund cleanup is similar." Cassie and GWPDX knew that a full cleanup would require a substantial fight. GWPDX also understood that without specific provisions to mitigate displacement, such as community-controlled development, community benefit agreements, and permanent affordable housing, Black and Indigenous communities, other communities of color, and lower-income white households were at risk of being fully excluded from inner-Portland neighborhoods. Grassroots groups would need to insert themselves into the cleanup process on their own terms in order to ensure a thorough remediation that actually benefited those who had already been displaced and others at risk of future displacement. Cassie began alerting representatives from other community organizations.

Cassie, Jeri, and a few dozen people soon gathered at a coffee shop in the back of the former Talking Drum Bookstore in Albina. Jeri describes the crowd: "We had a full room," recalling that leaders from Black, Indigenous, Latinx, Eastern European, and other groups attended. Analysis at Talking Drum painted a critical picture of the PHP's communications. "The general sentiment about those focus groups was kind of shock at how bad the contamination was—people didn't realize [the extent of] it. And concern, and that they wanted to do something about it, and that they didn't trust the PRPs to keep facilitating these outreach and engagement plans," explains one participant.

Crucially, the dozen or so mainly small grassroots organizations that came to form PHCC represented diverse backgrounds and geographies. Jeri offers a summary: "What was interesting was leaders from those communities [of color] came. The [larger] organization staff didn't come, but *leaders—the people on the ground*—came." Some were affiliated with smaller nonprofit organizations, whereas others were part of grassroots groups with no formal designations. Jeri reflects on the organic process of assembling such a wide variety of participants: "So [GWPDX] did organizing to bring all these groups together, to get people from all these different walks of life. And pretty soon you're bringing in . . . all these people [who] are going, 'Of course the river's the most important thing to us, of course water's important to us, of course having fish is a cultural thing for us and a spiritual thing for us.' So you brought together all of these people who historically have a much deeper understanding with the land than many [white] Americans do." For the groups that became PHCC's core members, the river that cuts through the heart of Portland was a central unifying concern, and indeed it ultimately became the element that stitched such disparate groups together to collaborate for years to come.

COMPLEX GEOGRAPHIES OF RESISTANCE

Many of the groups and individuals that made up PHCC's original base were motivated by concrete, metabolic, place-based attachments to the Portland Harbor—even if they did not live near the harbor. Several people relied on the river for basic sustenance. One participant, from the East European Coalition, explains, "[There are] a lot of [Eastern European] fishermen out there, who don't know about why you cannot fish there from the river; that's why we got involved." Members of Líderes Verdes, a Latinx-led EJ group based nine miles east of the harbor, also joined, because they were concerned about the health of community members who rely on the river to feed their families. "The water is very polluted. We nor our children nor our grandchildren are able to eat the fish. And that's why I like to participate," one Líderes Verdes member says.

A Black elder named Wilma Alcock, who later would sit on PHCC's Steering Committee, reflects fondly on her childhood eating fish from the Willamette, alongside hush puppies, potatoes, onions, and salad. Fish provided vital sustenance for Wilma's family, and the whole process of fishing and cooking was a source of leisure and comfort for her community in a time of intense racial discord during and after World War II:

> We fished all up and down the river, wherever there was a bank you could get to. . . . It was a pastime and a feeding. We ate fish probably once a week because it was just healthy for you. . . . And we also would be out at Sauvie Island and after I got older, [my parents] had a little trailer, it was called a Teardrop, and it would attach to the back of a car. There was a place for a stove, and a place to sleep, and they would catch fish out in the river. We would use clean water and clean [the fish] and use salt and pepper and cornmeal and fry it in that black cast iron skillet. Oh my god, there was nothing any better than that.[20]

When she joined PHCC, Wilma immediately made the connection that the harbor's health was intimately linked to that of her community.

Another PHCC member, Mike, from the houseless-led group Right 2 Survive, recalls growing up in Portland in the 1960s, cutting steelhead that his father caught in surrounding rivers into thick steaks. Adding slices of lemon, salt and pepper, and onion rings, his family "hobo-cooked" the fish over campfires, in aluminum foil packets. A concern for the safety of fellow houseless people living along the river today—some of whom are not aware of the toxic fish—drove Mike's participation in PHCC.

With the exception of a handful of houseless people, nearly no one who joined PHCC in its early days actually lived within walking distance of the harbor. For many, place attachment to the harbor, and in turn involvement in PHCC, revolved around a desire to safely access life-sustaining foods, regardless of proximity to the river.

Yet many PHCC members and their ancestors *used* to live near the harbor. Place attachment to the harbor is thus made more complicated by histories of dispossession and displacement, and these histories inspire attention to myriad issues beyond fish and pollution. For many in Portland's Black community, equitable jobs and housing, for instance, are inseparable from the health of the river, and this nexus of concerns motivates participation in PHCC. One person explains, "I am here to make sure that we facilitate and leverage this Superfund site into something that actually benefits the displaced people who have been suffering not only from the toxins, but from the job loss, from displacement in housing, education, employment, children, the whole nine yards." Overarching and interrelated issues in the harbor and beyond, including intergenerational health, workforce development, housing, and more, thus motivate involvement in the harbor cleanup, especially for those whose people no longer live near the harbor due to gentrifying forces.

In thinking back to the river of her childhood, Wilma further explains the linkage between the work of healing the water and healing her community, which is fragmented by displacement: "My hope is that [the river] gets back to where it once was where it can be life sustaining for people. Because really, actually, progress just—it doesn't sustain people, it just—well, ok, progress, if I'm talking about progress when you're displacing something and putting something else, it usually doesn't last." Contrasting the so-called progress of city building with the restoration of a landscape and waterway that could actually support human life, and in particular the health and spirit of the Black community, Wilma's statement encapsulates the collective thinking of many PHCC members: development, green or otherwise, is futile without simultaneous attention to life-giving elements of the landscape, such as the river. At the same time, river remediation is in vain if only white people of means steer the process and benefit from the outcome, especially if profits come at the expense of those who have suffered from pollution, displacement, land theft, and other violences (see figure 7.2).

For houseless people and others, the prospects of future displacement also motivated involvement. Right 2 Survive leaders know that throughout the city, police and contracted cleanup crews periodically force houseless people to pack up and move, often ostensibly motivated by environmental concerns. "Move to where?" Right 2 Survive members ask. Advocates fear that as the harbor cleanup gets underway, remediation will become a convenient excuse to conduct sweeps with even more urgency along the river. Right 2 Survive member Loretta Pascoe articulates the contradictions that people living along the river face: "People along the river need a safe place to go, and we need to be able to afford to get that [housing] so that we can be away from the river while the cleanup is [taking place]."

FIGURE 7.2. Portland Harbor Community Coalition members marching for racial justice. Photo by author.

Other complex spatial concerns were particularly key motivators for Indigenous groups joining PHCC. Leaders with the Portland chapter of the American Indian Movement (AIM) put recovery of the Willamette at the heart of their participation, even though AIM members are scattered across the region. AIM leads seasonal water ceremonies on the banks of the Willamette and nearby waterways. People travel from as far as Umatilla and Pendleton, Oregon, and Yakama, Washington, for these ceremonies, some following in the footsteps of ancestors who seasonally traveled downriver along the Columbia to the Willamette Valley. Others, such as members of the Purépecha community, originally from Michoacán, Mexico, and who now live in Woodburn and Hubbard, Oregon, often drive forty miles north to dance in AIM's ceremonies, hosted along the river. "People can be healed, rivers can be healed," succinctly explains Art McConville (Nimiipuu [Nez Perce]/Cayuse), a local AIM chapter leader and longtime PHCC member.[21]

Other members describe similarly complex attachments to the harbor as driving their participation. Hadi identifies connections between Portland's river and those of Iraq, his home country, as motivating his involvement:

> I remember in my country, we have two rivers: Tigris and the Euphrates. Old rivers, yes. They come from Turkish mountains, very clear water, very clean, and one of them passes into Syria lands, and passes into Iraq. When [the river] reaches the south of Iraq, it is polluted heavily. Many diseases. The water becomes bitter or salty, not good to drink, or not good to water even the plants. I make a comparison between the Tigris and Euphrates and Willamette. We are suffering the same thing. So I think they [should] take care, the government, if they warn people or notify them, "Please, you have to take care of the rivers," not for Portland only, for all American states.

Irina, a member of the East European Coalition, likewise became concerned about the Willamette River because of her experiences growing up on another continent. Speaking about the Neva River in Saint Petersburg, Russia, Irina says, "I watched the river of my birth city and how it was polluted. From age five or six to age twenty, I saw sand getting dirtier, water getting more oily. There were no warnings." The daily ritual of drinking tea and serving it to friends and relatives who came to visit, bringing tea as a gift from Uzbekistan, Kazakhstan, Tajikistan, and other places along the Silk Road to China, is what really brought the problem to the forefront for Irina. "The tea we drank as a little girl was really good," she recalls. "[Friends] brought really good green and caravan teas." By the time she was a teenager, however, the city's tap water—from the Neva—had taken a turn for the worse. Irina remembers, "I was embarrassed to give [our guests] tea. The water tasted terrible. . . . The water did not allow us to enjoy the flavors." Irina connects her experience with this deterioration in drinking water quality to her work with PHCC: "I do not want Portland to go this direction," she says.

For nearly all PHCC member groups, then, a medley of issues both near and far from the harbor, from jobs to housing, spiritual connection to sustenance, and intergenerational concerns to the health of the river itself, are woven throughout analysis of the cleanup. Despite PHCC's participants hailing from disparate geographies, the Willamette River has served as a convergence point. As a spiritually sustaining, "life-giving" element, the Willamette River—and rivers of the world—stitched PHCC's originating groups together. "Our River, Our Future" would become the coalition's tagline.

DIVERGENT UNDERSTANDINGS OF PLACE ATTACHMENT

With such diverse, complex geographies at the heart of its work, in 2013, PHCC leaders communicated with EPA officials about making the planning process more accessible to frontline communities. In mid-2014, EPA officials agreed to meet with PHCC leaders and to update the coalition on the status of a web-based tool the EPA had designed to help so-called EJ communities. The tool featured environmental and demographic data that would supposedly allow for patterns of impact to emerge as well as to correlate health outcomes with land uses. EPA officials stressed that the technology would allow users to compare trends happening in multiple places.

But PHCC representatives quickly realized that there were major shortcomings with the tool. They wondered how a snapshot of present-day contamination and demographics would help the coalition's advocacy, given PHCC members' complex historical ties to the harbor. Indigenous and Black communities, in particular, had been displaced over the course of decades and centuries and did not live in neatly organized census tracts near the river. And recent immigrants and refugees were dealing with cumulative impacts of toxins, food insecurity, housing instability, and more in their home countries and Portland alike, even as most lived far from the harbor. The EPA's tool would not reflect such complexities, coalition representatives stressed. Moreover, many people travel hundreds of miles to fish, canoe, and pray in the Portland Harbor. The mapping tool would not account for such a wide-ranging radius, nor would it capture houseless people living along the river today. PHCC representatives left the meeting, shaking their heads.

At issue was a discrepancy in the ways in which PHCC and EPA leaders understood peoples' place attachments. EPA officials conceptualized place attachment as legible through a map that would depict present-day proximity to the harbor. By and large, place attachment develops through spending time in a given place, whether a beautiful coastline, a particular neighborhood, or a nearby supermarket. For PHCC members, indeed, time is a key ingredient in many peoples' feelings of deep connection to the river and in their decision to fight for a cleaned-up harbor. Yet this time-oriented place attachment developed and motivated efforts to join PHCC despite physical separation, for many, from the harbor. Place attachment, then, can develop beyond one's tenure in a given place. A symbolic relationship, based not necessarily on a physical connection but rather on something less tangible, can contribute to shared meanings of a place. These shared meanings then become the basis of a group's efforts to defend and shape the future of a place.[22]

PHCC's work, to be sure, is rooted in the Portland Harbor as a real, physical space. But for many communities, their relationship to the harbor is simultaneously symbolic, relational, more-than-local. Such complex relationships to the harbor invoke narratives of violent displacements both toward and away from the harbor; memories of rivers of other times and places; and visions of a healthier, greener, more just future for the entire region and beyond. Put simply, relationships to particular places cannot always be neatly mapped or categorized. To truly embrace the needs and desires of frontline communities, it is necessary for policymakers, planners, researchers, the courts, and the general public to take seriously a more expansive conceptualization of place attachment. In the meantime, groups such as PHCC and others throughout Cascadia will continue to practice new ways of being in relationship to one another and to the landscape, in a collective effort to remake urban space for people rather than profit: Our River, Our Future.

NOTES

I gratefully acknowledge Portland Harbor Community Coalition members for allowing me to work and learn from and alongside them for many years. Appreciation is also due to Cassie Cohen and Jeri Jimenez for feedback on this chapter. Additionally, I am grateful to Donovan Smith for his collaboration in making an earlier version of a peoples' history of the Portland Harbor public, in the city's weekly *Street Roots*; to Steven Christian, who wrote and illustrated a comic book version of this peoples' history; and to PHCC members and Portland State University students, who helped produce the short film *A People's View of the Portland Harbor.* Portions of this chapter also appear in a paper published in *Antipode,* and I explore some of the issues presented here in a forthcoming book with the University of Georgia Press. This material is based upon work supported by National Science Foundation (NSF) IGERT Grant #0966376: "Sustaining Ecosystem Services to Support Rapidly Urbanizing Areas." Any opinions, findings, and conclusions or recommendations expressed here are those of the author and do not necessarily reflect the views of the NSF.

1 Lizzy Acker, "Ted Wheeler to Swim the Willamette Again," OregonLive, updated January 9, 2019, www.oregonlive.com/portland/index.ssf/2016/08/ted_wheeler_to_swim_the_willam.html.

2 Lonely Planet, "Best in Travel 2017: Top 10 Cities," October 25, 2016, www.lonelyplanet.com/articles/best-in-travel-2017-top-10-cities.

3 Quoted in Lizzy Acker, "We Swam across the Willamette with Ted Wheeler and It Wasn't Even Gross at All," OregonLive, updated January 9, 2019, www.oregonlive.com/trending/2016/08/we_swam_across_the_willamette.html.

4 Julian Agyeman, *Introducing Just Sustainabilities: Policy, Planning, and Practice* (New York: Zed Books, 2013).

5 Katherine McKittrick, *Demonic Grounds: Black Women and the Cartographies of Struggle* (Minneapolis: University of Minnesota Press, 2006), xiii.

6 Setha M. Low, "Symbolic Ties That Bind: Place Attachment in the Plaza," in *Place Attachment*, ed. Irwin Altman and Setha M. Low (New York: Plenum Press, 1992), 165–84.

7 Erin Goodling, "Urban Political Ecology from Below: Producing a 'Peoples' History' of the Portland Harbor," *Antipode* (2019), https://doi.org/10.1111/anti.12493. Portions of this chapter appear in Goodling, "Urban Political Ecology," as well as in Erin Goodling and Donovan M.

Smith, "Below the Surface: A People's History of Portland Harbor," *Street Roots*, July 13, 2018, https://news.streetroots.org/2018/07/13/below-surface-people-s-history-portland-harbor; and Portland Harbor Community Coalition (PHCC), *A People's View of the Portland Harbor*, film, posted July 20, 2015, www.youtube.com/watch?v=uVCpyrNbNx8.

8 PHCC, *People's View.*

9 David V. Ellis, "Cultural Geography of the Lower Columbia," in *Chinookan Peoples of the Lower Columbia*, ed. Robert T. Boyd, Kenneth M. Ames, and Tony A. Johnson (Seattle: University of Washington Press, 2013), 42–62; David G. Lewis, Eirik Thorsgard, and Chuck Williams, "Honoring Our *tilixam*: Chinookan People of Grand Ronde," in Boyd, Ames, and Johnson, *Chinookan Peoples*, 307–25; and Henry B. Zenk, Yvonne P. Hajda, and Robert T. Boyd, "Chinookan Villages of the Lower Columbia," *Oregon Historical Quarterly* 117, no. 1 (2016): 6–37.

10 D. Ann Trieu Gahr, "Ethnobiology: Nonfishing Subsistence and Production," in Boyd, Ames, and Johnson, *Chinookan Peoples*, 63–79; and Lewis, Thorsgard, and Williams, "Honoring Our *tilixam*."

11 Robert T. Boyd, "Lower Chinookan Disease and Demography," in Boyd, Ames, and Johnson, *Chinookan Peoples*, 229–49; David Lewis and Thomas Connolly, "White American Violence on Tribal Peoples on the Oregon Coast," *Oregon Historical Quarterly* 120, no. 4 (2019): 368–81; Ronald Spores, "Too Small a Place: The Removal of the Willamette Valley Indians, 1850–1856," *American Indian Quarterly* 17, no. 2 (1993): 171–91; Mark Axel Tveskov, "A 'Most Disastrous' Affair: The Battle of Hungry Hill, Historical Memory, and the Rogue River War," *Oregon Historical Quarterly* 118, no. 4 (2017): 649–51; and Gray H. Whaley, *Oregon and the Collapse of* Illahee: *U.S. Empire and the Transformation of an Indigenous World, 1792–1859* (Chapel Hill: University of North Carolina Press, 2010).

12 Kenneth R. Coleman, "'We'll All Start Even': White Egalitarianism and the Oregon Donation Land Claim Act," *Oregon Historical Quarterly* 120, no. 4 (2019): 414–37.

13 Ann Curry-Stevens, Amanda Cross-Hemmer, and Coalition of Communities of Color, *The Native American Community in Multnomah County: An Unsettling Profile* (Portland: Portland State University, 2011), http://pdxscholar.library.pdx.edu/cgi/viewcontent.cgi?article=1093&context=socwork_fac; Karen J. Gibson, "Bleeding Albina: A History of Community Disinvestment, 1940–2000," *Transforming Anthropology* 15, no. 1 (2007): 3–25; John Linder, "Liberty Ships and Jim Crow Shipyards: Racial Discrimination in Kaiser's Portland Shipyards, 1940–1945," *Oregon Historical Quarterly* 120, no. 4 (2019): 518–45; Rudy Pearson, "African Americans in Portland, Oregon, 1940–1950: Work and Living Conditions—A Social History" (PhD diss., Washington State University, 1996); Sandy Polishuk, "'They Can't Come in through the Front Door because You Guys Won't Let Them':

An Oral History of the Struggle to Admit African Americans into ILWU Local 8," *Oregon Historical Quarterly* 120, no. 4 (2019): 546–63; and Alonzo Smith and Quintard Taylor, "Racial Discrimination in the Workplace: A Study of Two West Coast Cities during the 1940s," *Journal of Ethnic Studies* 8, no. 1 (1980).

14 James V. Hillegas-Elting, *Speaking for the River: Confronting Pollution on the Willamette, 1920s–1970s* (Corvallis: Oregon State University Press, 2018).

15 Ellen Stroud, "Troubled Waters in Ecotopia: Environmental Racism in Portland, Oregon," *Radical History Review* 74 (1999): 65–95.

16 Erin Goodling, Jamaal Green, and Nathan McClintock, "Uneven Development of the Sustainable City: Shifting Capital in Portland, Oregon," *Urban Geography* 36, no. 4 (2015): 504–27.

17 Lisa K. Bates, "Growth without Displacement: A Test for Equity Planning in Portland," in *Advancing Equity Planning Now*, ed. Norman Krumholz and Kathryn Wertheim Hexter (Ithaca: Cornell University Press, 2019), 21–43.

18 Marisa A. Zapata et al., *Governance, Costs, and Revenue Raising to Address and Prevent Homelessness in the Portland Tri-County Region* (Portland: Portland State University, 2019), www.pdx.edu/homelessness/sites/g/files/znldhr1791/files/2020-05/RegionalHomelessnessReport.pdf.

19 Roxanne Dunbar-Ortiz, *An Indigenous Peoples' History of the United States* (Boston: Beacon Press, 2014).

20 PHCC, *People's View.*

21 PHCC, *People's View.*

22 Low, "Symbolic Ties That Bind."

CHAPTER 8

THE PROGRESSIVE PROMISE OF RECONCILIATION IN VANCOUVER'S NORTHEAST FALSE CREEK

GIUSEPPE TOLFO

IN 2018, THE CITY OF VANCOUVER RELEASED THE NORTHEAST False Creek (NEFC) Plan to guide the removal of the Georgia and Dunsmuir Viaducts, two pieces of highway infrastructure that segregate Vancouver's southeastern half from its northeastern half. To replace the viaducts, the NEFC Plan prescribes a vibrant neighborhood that will darn the urban fabric and reintroduce intimacy to the public realm. Recalling Vancouver's own definition of *Vancouverism*, the NEFC Plan is a road map to "a new kind of city living" that combines "respect for nature" with "busy, engaging, active streets and dynamic urban life."[1] But despite these consistencies with institutionalized best practices, the NEFC Plan opens with a bold new promise: "to reconnect through reconciliation, replacing the Georgia and Dunsmuir viaducts with strong cultural, social and physical linkages."[2]

Reconciliation is a recent focus in Vancouver, the by-product of Canada's Truth and Reconciliation Commission investigation into the state's use of residential schools as an attempt to eliminate the social, cultural, and religious identity of its First Nations peoples. The commission's report confirmed that the impact of these residential schools continues to affect First Nations communities in the present and reinforced an understanding of the devastating effects that colonization continues to have on Indigenous peoples across Canada. In response to this report, Vancouver positioned itself as a City of Reconciliation and put forward a framework to deepen cultural understanding and improve relationships with urban Indigenous and First Nations communities. Recognizing the lasting and devastating effects of colonialism—including racism, poverty, and poorer health outcomes in the present—the City of Vancouver also extends the promise of reconciliation toward all cultural communities negatively impacted by city policies.

This chapter is a reflection on and an evaluation of the promise "to reconnect through reconciliation." It begins by considering the history of the neighborhoods surrounding False Creek, highlighting instances for which Vancouver might appropriately seek reconciliation, and the relationship between these transgressions and the promise of a more sustainable and livable city. As I will show, planning efforts in and around False Creek, whether motivated by colonialism, modernist planning principles, or the ambiguous promise of livability, have consistently ignored the interest of those on the margins. As a result, planning efforts have consistently contributed to displacement.

In the light of this context, this chapter interrogates the presumed connection between the NEFC Plan and reconciliation. It shows that policy writers only recently began to consider reconciliation as a policy priority, influenced by Canada's broader commitment to Truth and Reconciliation. But while reconciliation is a recent priority, the specific policy directives proposed for NEFC have deep roots. As a result, the promise of reconciliation manifests as an extension of the City of Vancouver's long-standing commitment to livability—a commitment that has historically prioritized the interest of experience-seeking urbanites over those of marginalized groups. Reconciliation comes across as little more than a branding exercise for these creative city planning efforts, as window dressing rather than a point of inspiration or a means toward inclusive urban policy. The chapter therefore highlights a contradiction: the promise of reconciliation is made with specific reference to the displacement of cultural communities, but by failing to consider Vancouver's history of dispossession comprehensively, the NEFC Plan reinforces the systemic preconditions that perpetuate such displacements in the present.

Reconciliation is indeed a demanding and worthwhile city building objective. But in the context of an increasingly polarized twenty-first-century city, seeking reconciliation demands that cities consider all those who have not been offered equal opportunity to flourish. This chapter demonstrates that without a comprehensive commitment to addressing entrenched injustice and systemic inequality, the NEFC Plan fails to amount to meaningful reconciliation.

THE VIADUCTS AND A NEIGHBORHOOD LOST

The Georgia and Dunsmuir Viaducts are located at the southeastern edge of Vancouver's downtown core, at the meeting point of several residential and commercial districts. These neighborhoods tell the story of Vancouver's evolution, from resource processing powerhouse to sustainable metropolis. But as the city has found its place among the world's most sustainable and livable, the benefits have not been shared equally. Over the course of False Creek's history, the well-being of marginalized communities was sacrificed, and thus Vancouver must address many issues in seeking reconciliation.

Vancouver is located on the unceded territories of the Musqueam, Squamish, and the Tsleil-Waututh Nations (see chapter 5). The original Granville Townsite began as a makeshift extension of the Hasting's Sawmill, but the true success of the colonial project was borne of the Canadian Pacific Railway's 1885 decision to extend its western terminus stations past New Westminster, through False Creek, and into Coal Harbor.[3] Following this decision, over the next thirty years, the city's prosperity and growth was largely tied to production along the shores of False Creek and the Burrard Inlet.[4] The success of Vancouver's colonial project, measured both in terms of prosperity for the settlers and the erasure of First Nations' communities, is therefore tied to False Creek's evolution.

From little more than a marshland, False Creek became a major resource processing and distribution center, transforming a small logging community into a bustling commercial metropolis. By 1910, twenty industrial firms existed along the False Creek shore, and residential communities grew to sustain the industrial expansion: between 1901 and 1911, Vancouver's population quadrupled.[5] Such was the founding of Strathcona, Vancouver's oldest neighborhood of detached dwellings, located at the northeastern edge of False Creek.

When first constructed, Strathcona was home to a community of Italian laborers. As the community prospered, success drew its residents east and away from the industrial core—out of what was then known as the East End but is today the Downtown Eastside. The abandoned and vacant spaces in and around Strathcona welcomed later-stage immigrants—racialized Canadians who were otherwise excluded from the housing market.[6] Much of Strathcona filled up with Chinese Canadian residents, and for the better part of the first half of the twentieth century, the center of Black Canadian life in Vancouver became a T-intersection at the southwest end of Strathcona, known colloquially as Hogan's Alley. Small in absolute terms, the community represented a significant portion of Vancouver's Black population: chapels, churches, chicken houses, and speakeasies filled the area to serve a community growing in prominence.[7]

As only 1 percent of Vancouver's current population self-identifies as Black, it might come as a surprise that a prominent Black Canadian community once had a home in Vancouver's downtown core. But Hogan's Alley's forgotten eminence is the result of its transience. And Hogan's Alley's transience was enforced.

In 1928, the first of Vancouver's planning initiatives that were hostile toward the Black Canadian community was released. Ostensibly an effort to rationalize land use patterns, the Bartholomew Plan outlined distinct zones for residential development, for commercial activity, and for industrial production. In accordance with the plan, the City of Vancouver rezoned Hogan's Alley, abandoned by more affluent residents due to nearby industry, for factories, warehouses, and light industrial uses. Following the change, banks refused the neighborhood access to residential mortgage capital.[8] Hogan's Alley was redlined.[9] Repairs and upkeep fell. Industrial uses encroached further upon residential communities. By World War II, many of the area's homes and once-thriving commercial corridors were in various states of disrepair.

Conditions grew worse in the postwar years as structural shifts in the resource economy accelerated deindustrialization along the waterfront. With fewer working-class employment opportunities, communities and commercial districts spiraled further into decline. In Strathcona, the association between race, transience, and disinvestment triggered revanchist slum rhetoric.[10] Hogan's Alley was specifically considered a problem spot by the media, full of "squalor, immorality, and crime."[11] The presence of church services, once-thriving businesses, and families was easily forgotten when contrasted with harsh degradation imagery.[12]

It was in this unfortunate context that support for urban renewal found traction. Modernist urban renewal policies encouraged the expropriation and the removal of disinvested areas, replacing these often low-income and marginalized communities with highway infrastructure that would support the improved mobility of affluent suburban residents traveling to the commercial core. Facing decline in central city neighborhoods, and fearing the spread of "blight," this policy response grew in popularity across North America in the 1960s—and Vancouver was no exception. Inspired by modernist renewal policies, heavy-handed city engineers—cloaked behind a concern for efficiency and infrastructure provision—made "erasing the late-nineteenth-century character of Strathcona" and its surrounding neighborhoods official city policy and planned to replace much of Vancouver's East End with a waterfront freeway.[13]

Using mainstream media as an indicator of popular opinion, the City of Vancouver's renewal policies were widely praised.[14] But just as renewal was gaining traction among Vancouver's propertied elite, the 1960s also saw the rise of a reactionary reform movement, inspired by a growing resentment of modernist planning principles. To these reformers, the city was not a place of geometry and efficiency but a social space. Instead of harboring distaste for the uncontrollable urban experience, reformers celebrated the city's vibrancy and diversity.

In response to the waterfront freeway initiative, the Chinese Canadian community formed the Strathcona Property Owners and Tenants Association (SPOTA) and banded together with counterculture activists to affirm the value of Vancouver's established neighborhoods: their scale and character, the lifestyle associated with the central city, and the community of established residents providing central city spaces with life. Led by SPOTA, this diverse group of community activists and residents came together to form a livable city coalition that put a stop to the freeway. Though they did prevent Chinatown and Gastown from being demolished, the livable city coalition was too slow to save Hogan's Alley. In 1968, the neighborhood was leveled to make way for the Georgia and Dunsmuir Viaducts, the first stage of Vancouver's waterfront freeway network.[15]

Today, the viaducts stand in their place at the edge of Chinatown, reminders of displacement in a landscape founded on colonial dispossession. But while the viaducts may act as tangible reminders of the community lost, the construction of the viaducts is not entirely to blame for the destruction of Hogan's Alley. While eventually considered a problem spot, at its height in the early twentieth century, Hogan's Alley was a thriving center of Black Canadian life. But modernist planning policies created a self-fulfilling prophecy: by characterizing the neighborhood exclusively suitable for industrial production, residential mortgage capital was refused, creating an environment unsuitable for residential life. The residential decline these policies reinforced eventually served as motivation for the community's destruction.

Interpreting the viaducts as symbols of community destruction does not sufficiently recognize the destabilizing effect redlining had on the community. Restricting neighborhoods access to essential community resources—residential mortgage capital—was as damaging as urban renewal. In recognition of this perhaps subtler form of injustice, Hogan's Alley finds a place within a wider history of dominant class prioritization, whereby policy redirects minority groups to the margins and ensures that these marginalized groups are ostracized.[16] This introduces a second misreading of the area's history: the failure to account for subtle injustices that have destabilized other communities along False Creek. Just as it is a mistake to assume the viaducts are entirely to blame for the erasure of Hogan's Alley, it is a mistake to assume that the Black Canadian community and the First Nations communities before it are the only ones to have been displaced from False Creek. This fact is made abundantly clear by False Creek's contemporary shoreline—where residential towers line the waterfront and little remains of the industrial activity that only fifty years ago polluted its shores.

Deindustrialization impacted most postwar North American and European communities, but Vancouver actively encouraged its post-industrial turnover when, beginning in the late 1960s and early 1970s, the downtown core was transforming into a place for middle-class, service sector employment. Many of these "new middle-class" residents were members of the same livable city coalition that challenged the waterfront freeway. But while SPOTA's organizing efforts certainly began as an affirmation of the value of the existing cultural community, their success was, at least in part, due to the overlap with shared interests of middle-class urbanites. For those middle-class residents inspired by sustainability, increased automobile dependency was an affront to their social and environmental values. For others inspired by the aesthetics of central city life, heavy-handed urban renewal was a threat to their aesthetic preferences. After putting a stop to the freeway, these middle-class tastemakers shifted their focus to improving "livability" in False Creek and across the city.[17]

Encouraged by the call of the middle-class environmentalist and gentrifier for improved livability, Vancouver City Council facilitated an accelerated decline of the resource-processing industries in False Creek, prematurely forcing working-class employment out of the city.[18] An award-winning residential community replaced the resource-processing industry at False Creek South, with social equity considerations—a mixture of innovative housing types and forms at a range of affordability levels—incorporated into the "livable" design.[19] But once the industrial shoreline was remediated and replaced with desirable community amenities, Fairview Slopes, a working-class neighborhood south of the redeveloped False Creek shoreline, became a hotbed for gentrification. At Fairview, the industrial history and working-class character were appropriated by postmodernism, branded as authenticity, and transformed into a consumable product.[20]

Following deindustrialization, False Creek's northern waterfront hosted the Expo 86 World's Fair. True to the entrepreneurial spirit of such festivals, Expo 86 introduced Vancouver to the world—as a unique beauty, a site for tourism, and a home for international investment capital.[21] The effects were felt immediately among Vancouver's marginalized populations as many of the single-room occupancy hotels across the Downtown Eastside, homes of last resort for the city's lowest-income residents and a substantial number of the city's urban Indigenous people, evicted their long-term tenants to capitalize on visiting tourists.[22]

After the fair, the Expo grounds were sold to Li Ka-shing, Hong Kong's wealthiest property developer, a sale that reinforced the globalization of Vancouver's real estate market.[23] The agreement of sale stipulated community amenity requirements for development of the Expo lands—including the provision of parks, schools, libraries, and other public spaces—but the affordable housing requirements were reduced from an originally proposed 33 percent to 20 percent.[24]

Despite the reduced commitment to affordability, Concord Pacific—the shell company used to develop the Expo lands—encouraged the association between the new high-rise condominiums and sustainability, appropriating False Creek's legacy without as significant a commitment to equity.[25] Lower-income, working-class, and marginalized communities benefited far less from this new form of "sustainability as density."[26]

As can be seen, the prioritization of dominant class interests existed long before the construction of the Georgia and Dunsmuir Viaducts and continues to effect marginalized residents in the neighborhood surrounding False Creek. The consequent displacement remained a threat in the lives of Vancouver's marginalized residents long after Hogan's Alley was erased from the urban landscape. In fact, to this day, Vancouver's marginalized residents are threatened by displacement in neighborhoods adjacent to False Creek. Most recently, gentrification has shepherded low-income residents into and out of the Downtown Eastside.[27] Vancouver's complicated history of injustice toward marginalized groups therefore extends into the present. If Vancouver intends to meaningfully reconcile for the history of Northeast False Creek, it has to contend with the systematic preconditions that allow marginalized communities to be lost and sacrificed: the prioritization of certain groups' quality of life—either enforced by policy instruments or by blind faith in the market—over the well-being of others.

THE POLICIES IN NORTHEAST FALSE CREEK

The NEFC Plan is the capstone on thirty years of planning efforts in and around False Creek. What follows is a summary of these planning efforts, contextualized by a close reading of relevant policy language, particularly of those documents that encourage viaduct removal. Overall, this summary and analysis demonstrate a lack of political interest in improving the lives of marginalized residents. Initially, planning documents ignored the interests and the experience of marginalized communities. Later, the experiences and history of marginalized communities were used to justify pre-established policy priorities. Across the entire period, neither reconciliation nor historic injustice had been a motivating force.

In 1990, Vancouver approved the False Creek North Official Development Plan (FCNODP) and with it a framework to develop a new mixed-use neighborhood along the northern shore of False Creek. The FCNODP provided little in the way of historic context for the area, but the historic "character of Yaletown" was praised for its "visual amenity."[28] False Creek's industrial history was recognized exclusively as it conferred dominant class benefit: ignoring post-industrial decline and the loss of working-class employment but referencing an aesthetic preference for adaptive reuse.

Overall, the FCNODP supported neighborhood connectivity in the place of adjacency and encouraged "strong visual and physical connections."[29] Most of FCN was sequestered for high-rise residential, but the area presently known as NEFC was to be made "identifiable, memorable, and lively" by enhancing the entertainment functions of the legacy resources leftover after Expo 86.[30] Residential use was considered at odds with this festival focus, and much of NEFC was set aside for commercial development.

By the mid-2000s, roughly ten million square feet of market residential had been constructed along FCN. The remaining permissions were those reserved in NEFC for commercial development. To appease restless landowners—"three of the four landowners . . . made proposals seeking significant residential development rather than the commercial potential"—a high-level review was announced by Council to investigate NEFC's residential development potential.[31] The resulting publication reversed City Council's earlier stance regarding condominium development in NEFC and proposed a new "highly urban, entertainment-oriented" form of livability, defined by the vibrant mix of uses: "The area will offer a different kind of livability that differentiates it from other waterfront areas that are predominantly places to live with few opportunities to work and play."[32]

At this point, NEFC was characterized as "vacant and underutilized."[33] Without a tangible expression of its history, apart from its entertainment assets, it was considered a blank slate. Starting afresh, in 2009, Vancouver City Council approved nine Guiding Principles for neighborhood development—emphasizing sustainability and the role of the area as a hub of activity and reinforcing connections between neighborhoods and between the city and the waterfront:

Creating a Special Destination in Vancouver
Establishing a Unique, Mixed-Use Neighborhood
Advancing Sustainable Community Development

Completing the False Creek Basin
Developing a Sense of Place
Encouraging Innovative Architecture and Variety in Building Form
Connecting to the Downtown
Fostering a Vibrant Waterfront
Providing a Connected Active Public Realm

These development objectives were similar to those guiding development since the 1970s.[34] Replete with city building catchphrases such as "destination, "unique," "sense of place," and "vibrant," these development guidelines were also similar to those that support creative city building: an economic development strategy that leverages neighborhood aesthetics as amenity to attract "creative" professionals.[35] As creative city strategies have been criticized for prioritizing the preferences of gentrifiers and discounting the experience of marginalized residents, one could reasonably expect that a neighborhood created under the guidance of these principles will inadequately address systemic inequality.[36]

By 2010, maintaining the Georgia and Dunsmuir Viaducts was growing prohibitively expensive, and viaduct removal became a subject of intense debate. Vancouver City Council commissioned a study to investigate retention, remediation, and removal scenarios. But, as if to ensure the strategy for any lands released aligned with the Guiding Principles in NEFC, the summary documents leveraged history as a malleable force of justification.

The 2011 *Viaducts and False Creek Flats Planning Report* referenced the once-proposed freeway network, but it did not mention Hogan's Alley. Instead, it compared the original function of the viaducts—a connection between the downtown core and its adjacent residential neighborhoods—with the contemporary structures' impacts on neighborhood continuity: "The first Georgia viaduct, completed in 1915, was built to connect the eastern part of the City to the Downtown core. The railway yards and industrial area below created a barrier and the viaduct created a way to remove this barrier with a relatively flat connection to the bluff of the downtown. . . . Today [the viaducts] create a *barrier* for the area below and for linkages between Chinatown, Gastown and Strathcona to False Creek."[37]

Following the report, the council commissioned a transportation study to investigate the effects of viaduct removal. The transportation study report described how public response to the freeway plans in the 1960s acted as "catalyst for change," encouraging a new form of urban planning, eventually contributing to Vancouver's regeneration: "The emphasis on accommodating the automobile waned, and a new approach to urban planning was developed in the 1980s that recognized the downtown area as a place for people to live. . . . The redevelopment of Downtown South and the former industrial lands at False Creek North and Coal Harbor for primarily high-density residential use has reinforced this transformation. . . . The renovation of heritage buildings in Yaletown and Gastown . . . have further contributed to this regeneration."[38] Unlike the earlier planning report, this transportation study recognized that urban renewal previously contributed to the eviction of a "predominantly African-Canadian community known as Hogan's Alley."[39] But neighborhood revitalization and livability remained the focus and historic injustice an afterthought rather than a motivation. That the lived experience of former Hogan's Alley residents was not a primary motivator is reinforced by the fact that the report fails to consider the similar experiences of the other marginalized residents displaced from the city.

According to the transportation study, the viaducts are reminders of a time before livability was prioritized and continue to constrain its achievement: "Legacies of past transportation planning decisions continue to inhibit the achievement of current goals regarding livability and sustainable transportation. . . . In Vancouver, this is exemplified by the provision of the Georgia and Dunsmuir viaducts."[40] This narrative does not consider that False Creek had once been a center for working-class employment and that current high-density condominiums do not serve the same demographic group. It does not consider how heritage revitalization contributed to gentrification.[41] Overall, the benefits of revitalization are praised, but the experiences of marginalized residents erased.

In 2013, a policy report regarding the viaducts was presented to City Council and reinforced earlier characterization of the viaducts as mistakes limiting Vancouver's livability: "In every city's evolution there are rare opportunities to take bold city building steps to advance the city's goals and livability or correct a past planning wrong. The potential removal of the viaducts provides an opportunity for the City of Vancouver to do both."[42] In this report there are several references made to Hogan's Alley as a distinct place but only one reference to the Black Canadian community displaced. Without reference to the neighborhood's residents, history is reimagined to benefit future consumption-seeking residents: "Forty years ago, buildings on the blocks between Quebec and Gore were demolished to make way for the viaduct structures. Removal of the viaducts allows for restoration of shops and services along the Main Street corridor, and the mixed-use development of the two city-owned blocks."[43] Here, the experience of dispossession is not of primary interest. The focus is on the aspects of the lost neighborhood that confer creative-class benefit: the reintroduction of shops and amenities. The history of the space becomes an identifier for the future neighborhood, providing it with a desirable brand. The viaducts are framed as barriers toward neighborhood connectivity, not as symbols of displacement; they are framed as opportunities for recreation rather than neighborhood re-creation.

From the outset, viaduct removal analysis was heavily influenced by preexisting policy in NEFC. As a result, any space made available by viaduct removal was conceived entirely in terms of its ability to procure creative-class serving amenities. Early policy documents failed to consider the experiences of marginalized residents outright. Later, when references to Hogan's Alley are made, there is little acknowledgment of the residents lost from the area, and commemoration efforts are similarly influenced by the Guiding Principles in NEFC: the City of Vancouver hopes to replace the viaducts with a unique "destination," recapturing the "vibrancy" of the blocks destroyed. Addressing a history of displacement, or reconciling with the cultural community lost, is not a priority. Nor are the experiences of those marginalized a prominent consideration: how the lands made available by viaduct removal might benefit low-income and marginalized residents—residents who share the experience of being displaced with previous Hogan's Alley residents—remains unconsidered.

THE NORTHEAST FALSE CREEK PLAN

Whereas earlier planning efforts in False Creek rarely acknowledged the Black Canadian community presence in Vancouver, the NEFC Plan is forthcoming about this history. But reconciliation—either with the original Coast Salish communities that had traditionally lived on lands today known as Vancouver or with the historic Black Canadian community of Hogan's Alley—was not the motivating force behind the NEFC Plan policy initiatives. While an interest in reconciliation may have inspired thorough engagement with history, the policy initiatives remain heavily influenced by the earlier Guiding Principles.

To understand how reconciliation became a topic of interest among Vancouver's policy writers, the lens must be widened to reference countrywide conversations. In 2009, Canada's Truth and Reconciliation Commission began a six-year-long effort to comprehensively understand the effects of Canada's residential school system on Canada's First Nations. In the summary report *Honoring the Truth, Reconciling for the Future* (or the Summary Report), the commission describes Canada's calculated efforts to eliminate the social, cultural, and religious identity of First Nations peoples.

The Summary Report recognizes its findings as fundamentally at odds with the self-proclaimed Canadian identity: a country that prides itself on its inclusivity and kindness. But the Summary Report also comments that while the truth may challenge and destabilize, achieving meaningful reconciliation is the greater challenge. It requires not just the recognition of past mistakes but a common understanding of the detrimental effects of the past in the present.

To align with Canada's efforts, in 2013, the City of Vancouver began a year of reconciliation and committed to advance greater understanding of the historic injustices that have shaped the lives and experiences of Indigenous people. On July 8, 2014, furthering its commitment to redress, City Council adopted a Framework for Reconciliation, and Vancouver was designated the first City of Reconciliation. City departments promised to embed reconciliation within their core work, and in 2018, the City of Vancouver expanded the original focus to include other cultural minorities. The NEFC Plan was released during this gust of turbulence and fanfare. Its bold promise to reconnect through reconciliation finds inspiration in Canada's efforts to address the physical, biological, and cultural genocide exacted on Indigenous communities.

With the NEFC Plan, Vancouver formally recognizes its role in having evicted the Musqueam, Squamish, and Tsleil-Waututh from their traditional territories and the Black Canadian community from Hogan's Alley. Aligning with this promise, representatives of First

Nations communities and urban Indigenous residents, the Historical Discrimination Against Chinese People in Vancouver Advisory Group, and the Hogan's Alley Working Group were consulted when preparing the NEFC Plan.

Within the NEFC Plan, the Hogan's Alley Working Group describes the removal of the viaducts as an opportunity to address the city's history of anti-Black racism: "[The] vibrant Black presence at Hogan's Alley was ultimately shattered by political planning schemes, first *slum clearance* and later *urban renewal*, each regarded by Black residents as methods of what was sardonically called *Negro removal*."[44] In line with this logic, the Hogan's Alley Working Group sees the viaducts' replacement as an opportunity for reconciliation: "With the viaducts coming down as part of this new NEFC Plan, a unique opportunity exists to bring a Black community presence and influence back to this area."[45]

Likely the result of the City of Vancouver's focused engagement with cultural communities, provisions are included in the NEFC Plan to ensure that Indigenous and Chinese Canadian cultural history are incorporated in place-names, place-keeping, and the design of public art and parks space. The NEFC Plan also promises First Nations and urban Indigenous peoples with direct canoe access to shore. The new Hogan's Alley Community Center onsite, the block plan, and the scale of the development that replaces the viaducts will all speak to the once-vibrant Black Canadian community.

These initiatives reinforce Vancouver's cultural communities' place in space and are promising steps in a progressive direction. But whether these initiatives qualify as reconciliation is worth consideration. Stephanie Allen's work with the Hogan's Alley Working Group confirms this: confined to an engagement process designed to limit their influence, demands for restorative justice and land held in trust were sidestepped and ignored by a city more interested in incorporating the Black Canadian influence in the design and programming of public space.[46] This is yet another example of how, under neoliberalism, representatives of marginalized groups have been forced to dull their more radical persuasions and demands to sustain even a minor influence.[47]

Testament to this diluted influence, some of the NEFC Plan initiatives toward reconciliation are heavily influenced by earlier policy initiatives: direct canoe access is promised, but a public boat launch was already proposed for Northeast False Creek.[48] Others are little more than reconciliation headdress on preexisting priorities: public art, low-density character areas, vibrancy, community amenities. Even the more significant initiatives, those that seem to address systemic poverty in the surrounding neighborhoods—neighborhoods within which urban Indigenous are disproportionately represented—can be traced to preexisting planning initiatives. The NEFC Plan requires that three hundred units of social housing be developed. This is approximately 30 percent of the total units that will be built to replace the viaducts. However, this same share of social housing was previously promised within the Downtown Eastside Plan and, under Vancouver's definition of "social housing," translates to only forty-five units at below-market rents.[49] With this additional context, it is worth considering whether marginalized communities' representation, culturally sensitive place-making, and increased cultural fluency are the limits of Vancouver's responsibility in seeking reconciliation.

As described in the Summary Report, challenging though it may be to recognize a need for it, meaningful reconciliation requires a break from the historic conditions that reinforce injustice. It requires more than a historic focus. It demands a restorative focus: a dedication to justice and equity. As the policies put forward in the NEFC Plan can be traced back to earlier initiatives, it cannot be considered a break from the past. While reconciliation is described as a "foundational component" of the NEFC Plan, it has instead been used to reframe preexisting policy initiatives.[50]

The NEFC Plan may well be an example of progressive city building, but considering it an example of meaningful reconciliation sets an underwhelming precedent. At the very least, the question of how a city might repair historic violence demands that reconciliation act as primary motivation. At its utmost, in the context of appropriation and erasure, reconciliation demands economic reparations.

The violence against cultural minorities in NEFC extends beyond enforced silence. The communities dispossessed lost more than a space to engage in cultural practices. These dispossessions are instances when economic autonomy and opportunities were stolen. While the NEFC Plan may have elevated the discourse of reconciliation and forced a conversation about historic injustice, the language and promise of reconciliation should do more than inspire understanding. Reintroducing cultural histories to a space, protecting cultural heritage, and improving cultural communities' visibility in the public realm will not address the economic consequences and the vulnerability disproportionately experienced by Vancouver's First Nations, urban Indigenous peoples, and cultural minorities. If Vancouver is to be successful in its efforts to repair the wound of stolen lands and livelihoods, it must be prepared to extend economic opportunity and spaces for prosperity to those from which these have been stolen.

The case for economic reparations is only reinforced when one considers ongoing experiences of gentrification and dispossession throughout Vancouver and the marginalized communities across Cascadia that are struggling with increased income and wealth inequality. Especially so if we contemplate the likely effect that the NEFC Plan—a city building strategy that will replace the viaducts with a vibrant and desirable community—will have on neighboring communities. The viaducts are relics of modernist city building that cause negative externalities for those in neighboring areas. As a result, they have dulled some of the gentrification pressures threatening Chinatown and the Downtown Eastside.[51] Replacing the viaducts with amenities and reintroducing waterfront access will attract even more affluent Vancouverites into the area. From this perspective, those who stand to benefit most from Vancouver's efforts to achieve so-called reconciliation at NEFC are in a similar class position to those who benefited from earlier rounds of appropriation and displacement at Hogan's Alley: dominant classes demanding convenient access to the city. As a result, such reconciliation may well find its place among a list of progressive priorities that have contributed to displacement for low-income residents in Vancouver.

"The Northeast False Creek Plan provides an opportunity to reconnect through reconciliation, replacing the Georgia and Dunsmuir viaducts with strong cultural, social and physical linkages."[52] To the extent that this introduction to the NEFC Plan promises to connect Vancouver to the waterfront and its history, it is a resounding success. To the extent that the NEFC Plan comprehensively addresses the prioritization of certain interests in policy agendas, or to the extent that it repairs the effects of injustice or violence against marginalized communities, the NEFC Plan barely scratches the surface. As a result, it comes across as performative. The progressive promise of reconciliation is used to brand the more substantial commitments to equity, as if to add political palatability to the policy agenda. The lofty rhetoric and the promise for reconciliation is a veil, concealing the fact that deep reparative measures are currently inconceivable. Vancouver's institutionalized focus on dominant class interests continues to this day and by extension pollutes Vancouver's reconciliation efforts in False Creek. So while Vancouver talks the talk, without a foundational commitment to economic reparations, the NEFC Plan is little more than business as usual.

NOTES

1 City of Vancouver, "Urban Planning, Sustainable Zoning, and Development," updated 2020, https://vancouver.ca/home-property-development/planning-zoning-development.aspx.
2 City of Vancouver, *Northeast False Creek Plan* (Vancouver: City Planning Department, 2018), 3, https://vancouver.ca/files/cov/northeast-false-creek-plan.pdf.
3 Nicholas Blomley, *Unsettling the City: Urban Land and the Politics of Property* (New York: Routledge, 2004).
4 Robert K. Burkinshaw, *False Creek: History, Images, and Research Sources Occasional Paper no. 2* (Vancouver: City of Vancouver Archives, 1984).
5 Burkinshaw.
6 Kay J. Anderson, *Vancouver's Chinatown: Racial Discourse in Canada, 1875–1980* (Quebec: McGill-Queen's University Press, 1991).
7 Wayde Compton, *After Canaan: Essays on Race, Writing, and Region* (Vancouver: Arsenal Pulp Press, 2010).
8 Compton, *After Canaan*.
9 See Richard Rothstein, *The Color of Law: A Forgotten History of How Our Government Segregated America* (New York: Liveright Publishing Corporation, 2017).
10 Anderson, *Vancouver's Chinatown*.
11 Compton, *After Canaan*, 91.
12 Rhodi Windson Liscombe, "A Study in Modern(ist) Urbanism: Planning Vancouver, 1945–1965," *Urban History* 38, no. 1 (2011): 124–49.
13 Mike Harcourt, Ken Cameron, with Sean Rossiter, *City Making in Paradise: Nine Decisions That Saved Greater Vancouver's Livability* (Madeira Park, BC: Douglas & McIntyre, 2009), 39.
14 Harcourt, Cameron, and Rossiter.
15 Compton, *After Canaan*.
16 See Blomley, *Unsettling the City*; Rothstein, *Color of Law*; Ted Rutland, *Displacing Blackness: Planning, Power, and Race in Twentieth-Century Halifax* (Toronto: University of Toronto Press, 2018); and Jeff Sommers, "Men at the Margin: Masculinity and Space in Downtown Vancouver, 1950–1986," *Urban Geography* 19, no. 4 (1998): 287–310.

17 David Ley, *The New Middle Class and the Remaking of the Central City* (New York: Oxford University Press, 1996).

18 Ley *The New Middle Class*; Jacopo Miro, "Visions of False Creek: Urban Development and Industrial Decline in Vancouver, 1960–1980" (master's thesis, University of Victoria, 2009), http://hdl.handle.net/1828/3513.

19 John Punter, *The Vancouver Achievement* (Vancouver: UBC Press, 2003).

20 C. A. Mills, "'Life on the Upslope': The Postmodern Landscape of Gentrification," *Environment and Planning D: Society and Space* 6, no. 2 (1988): 169–90.

21 David Harvey, "From Managerialism to Entrepreneurialism: The Transformation in Urban Governance in Late Capitalism," *Geografiska Annaler: Series B, Human Geography* 71, no. 1 (1989): 3–17.

22 Jeff Sommers and Nick Blomley, "The Worst Block in Vancouver," in *Stan Douglas: Every Building on 100 West Hastings*, ed. Reid Shier (Vancouver: Arsenal Pulp Press, 2002), 18–58.

23 Kris Olds, "Globalization and Urban Change: Tales from Vancouver via Hong Kong," *Urban Geography* 19, no. 4 (1998): 360–85.

24 Punter, *Vancouver Achievement*.

25 Noah Quastel, "Political Ecologies of Gentrification," *Urban Geography* 30, no. 7 (2009): 694–725.

26 Noah Quastel, Markus Moos, and Nicholas Lynch, "Sustainability-as-Density and the Return of the Social: The Case of Vancouver, British Columbia," *Urban Geography* 33, no. 7 (2012): 1055–84.

27 Blomley, *Unsettling the City*; Heather A. Smith, "Planning, Policy and Polarisation in Vancouver's Downtown Eastside," *Tijdschrift voor Economische en Sociale Geografie* 94, no. 4 (2003): 496–509; Katherine Burnett, "Commodifying Poverty: Gentrification and Consumption in Vancouver Downtown Eastside," *Urban Geography* 35, no. 2 (2014): 157–76; and Zachary Hyde, "Omnivorous Gentrification: Restaurant Reviews and Neighborhood Change in the Downtown Eastside of Vancouver," *City & Community* 13, no. 4 (2014): 341–59.

28 City of Vancouver, *False Creek North Official Development Plan* (Vancouver: City Planning Department, 1990), 16, https://vancouver.ca/files/cov/false-creek-north-official-development-plan.pdf.

29 Vancouver, *False Creek North Official Development Plan*, 5.

30 Vancouver, *False Creek North Official Development Plan*, 5.

31 City of Vancouver, *Policy Report—Development and Building: Removal of the Georgia and Dunsmuir Viaducts* (Vancouver: City Planning Department, 2015), 4, https://council.vancouver.ca/20151020/documents/rr1.pdf.

32 City of Vancouver, *Northeast False Creek: Directions for the Future* (Vancouver: City Planning Department, 2009), 11, https://vancouver.ca/files/cov/northeast-false-creek-directions-for-the-future.pdf; Vancouver, *Directions for the Future*, 16.

33 Vancouver, *Northeast False Creek*, 8.
34 Punter, *Vancouver Achievement.*
35 Richard Florida, *The Rise of the Creative Class* (New York: Basic Books, 2002).
36 See Jamie Peck, "Struggling with the Creative Class," *International Journal of Urban and Regional Research* 29, no. 4 (2005): 740–70; see Rutland, *Displacing Blackness.*
37 City of Vancouver, *Administrative Report—Viaducts and False Creek Flats Planning: Eastern Core Strategy* (Vancouver: City Planning Department, 2011), 8, emphasis added, https://vancouver.ca/docs/eastern-core/core-strategy-council-report.pdf.
38 City of Vancouver, *Summary Report: Vancouver Georgia and Dunsmuir Viaduct Study* (Vancouver: City Planning Department, 2011), 12–13, https://fraseropolis.files.wordpress.com/2017/04/2011-viaducts-study-summary.pdf.
39 Vancouver, *Summary Report*, 14.
40 Vancouver, *Summary Report*, 12.
41 Smith, "Planning."
42 Vancouver, *Policy Report*, 2.
43 Vancouver, *Policy Report*, 39.
44 Vancouver, *Northeast False Creek Plan*, 23, emphasis in original.
45 Vancouver, *Northeast False Creek Plan*, 23.
46 Stephanie Allen, "Fight the Power: Redressing Displacement and Building a Just City for Black Lives in Vancouver" (master's thesis, Simon Fraser University, 2019), https://summit.sfu.ca/item/19420.
47 Philip DePaolo and Sylvia Morse, "Williamsburg: Zoning Out Latinos," in *Zoned Out! Race, Displacement, and City Planning in New York City*, ed. Tom Angotti and Sylvia Morse (New York: Terreform, 2016), 72–95.
48 Vancouver, *Urban Structure.*
49 City of Vancouver, *Downtown Eastside Plan* (Vancouver: City Planning Department, 2015), https://vancouver.ca/files/cov/downtown-eastside-plan.pdf; Allen, "Fight the Power."
50 Vancouver, *Northeast False Creek Plan*, 20.
51 See David Ley and Cory Dobson, "Are There Limits to Gentrification? The Contexts of Impeded Gentrification in Vancouver," *Urban Studies* 45, no. 12 (2008): 2471–98.
52 Vancouver, *Northeast False Creek Plan*, 3.

PART 3

GOVERNANCE

CHAPTER 9

AGAINST "SEATTLE-IZATION"

Housing Justice and Activism in the Age of Amazon

JENNIFER L. RICE

> We shouldn't be inviting bullies to our neighborhood
>
> —ALEXANDRIA OCASIO-CORTEZ ON AMAZON HQ2

AMAZON.COM'S NOW INFAMOUS SEARCH FOR "HQ2" (A SECond corporate headquarters location outside of Seattle, Washington) received international attention from 2017 to 2019. The company released its HQ2 request for proposal (RFP) in September of 2017, and the eight-page announcement clearly touted the economic benefits of hosting HQ2: "Amazon will hire as many as 50,000 new full-time employees with an average annual total compensation exceeding $100,000 over the next ten to fifteen years. . . . The Project is expected to have over $5 billion in capital expenditures."[1] The RFP specified the company's preferred characteristics for its HQ2, which included an urban area of at least one million people, a "business friendly" environment, communities that "think big," and proximity to mass transit. The RFP also described Amazon's positive impacts in Seattle: "Amazon has been a catalyst for development in downtown Seattle with an abundance of restaurants, services, coffee shops, and for the redevelopment of South Lake Union and Denny Triangle with its sustainable buildings and open spaces." Cities that wanted to compete for HQ2 were required to outline available incentives, including "land, site preparation, tax credits/exemptions, relocation grants, workforce grants, utility incentives/grants, permitting, and fee reductions," reflecting the company's unabashed request that local governments fork over money, infrastructure, and resources to the then world's richest man (Amazon founder Jeff Bezos) and most valuable brand. Arguably, the HQ2 headquarters competition was a self-indulgent publicity stunt that could be orchestrated by no other company than Amazon.

Unsurprisingly, the RFP generated significant interest, and Amazon received 238 proposals from major cities, such as Boston, Toronto, and Atlanta, but also smaller communities, such as Greenville-Spartanburg, South Carolina, and Tulsa, Oklahoma.[2] In January 2018, the company released a list of twenty finalists and, ultimately, decided to share HQ2 between two locations: Crystal City, Virginia, and Long Island City, New York. Intense grassroots protests against HQ2 ultimately led the company to pull out of Long Island City (more on this later), but the company is now in the process of establishing tens of thousands of square feet of new office space in the Crystal City suburb of Washington, DC.

I discuss the HQ2 search at length because it shows the unprecedented international influence of Amazon on city governments, local economies, and urban communities. While most corporate headquarters location searches are only of interest to local and state officials concerned with "economic development," the HQ2 search was a nationally riveting, and infuriating, public spectacle, carefully devised by Amazon. It repeated the well-worn grand bargain with the entrepreneurial, neoliberal city that Amazon took to a new level: have a city provide some tax breaks, subsidize new infrastructure, and do some strategic marketing in exchange for jobs, jobs, jobs—a dance Seattle has been doing with Amazon for a decade.

This chapter examines the role of Amazon—one of the largest technology companies in the world—on urbanization and urban governance in Seattle and beyond. In particular, the chapter will examine four areas: (1) the rise of "Seattle-ization" as a means to capture the negative impacts of Amazon on the city, especially in the housing sector; (2) conceptual and theoretical tools that allow us to understand how and why Amazon is capable of influencing Seattle's housing market to such a degree; (3) the ways that activist groups in Seattle are challenging this system; and (4) how Seattle-ization is occurring, and being resisted, beyond Seattle and the Cascadia region. Given that Seattle is seen as a leader in social justice and environmental sustainability, a careful analysis of the role of Amazon in the city is helpful for understanding the impact of the technology sector on cities throughout North America. The goal here is to examine some of the processes that contribute to the negative aspects of Seattle-ization, namely the real estate state and settler colonial urbanization, and how we might resist them.[3]

Ultimately, the significant influence that companies such as Amazon have over our cities requires direct and significant confrontation. This is especially important as housing has become unaffordable in many urban areas, particularly in major coastal cities, such as Seattle. Inspired by those groups already working on the ground to challenge our current assumptions about housing and cities, this chapter calls for a radical rethinking of property relations to resist settler colonial urbanization. I will describe how this allows us to also reject the real estate state, which is driven by the interest of private developers and corporations who are some of the only ones that benefit from the new age of Amazon.

SEATTLE-IZATION: THE VANISHING CITY?

> It turns out other cities have been worried about becoming the next Seattle. That's right. "Seattle-ization" is a thing now. . . . It's described as "a particularly dire diagnosis," and characterized by high housing costs and tech wealth, a combination that's thoroughly transformed the city in an alarmingly short span of time.
>
> —GENE BALK,
> "'Seattle-ization'? American Cities Fear What's Happened Here"

Before there was an HQ2 search, Amazon's dramatic influence on its first headquarters location, Seattle, Washington, was receiving increased attention and scrutiny. Not long after Amazon started building its new corporate buildings in downtown Seattle in 2007, local media and advocacy groups noticed a rapid and troubling trend: the downtown neighborhood where the headquarters is located, South Lake Union (SLU), was showing classic signs of gentrification. A neighborhood that had previously been a "run-down" industrial and warehouse district was seeing the rise of tall, shiny new buildings, high-end grocery stores, doggy day cares, streetcars, and bike lanes.[4] Young tech workers with backpacks and blue Amazon badges around their necks started to fill downtown sidewalks, buses, bars, and restaurants. While concentrated in SLU, these trends quickly extended throughout downtown to the adjacent neighborhoods of First Hill and Capitol Hill and even the International District and Central District a few miles away (historically Asian and African American neighborhoods, respectively).

The demographic changes associated with the growth of Amazon's corporate headquarters in Seattle have been well documented. It has been widely noted that the young tech employees who work at Amazon want to live both an urban lifestyle and near their job, which influenced Amazon's choice to locate in downtown Seattle, as opposed to suburban Redmond, Washington, where Microsoft is located.[5] Most of the people who work at Amazon move to Seattle from outside of the city, and at the time of this writing, it was reported that more than one thousand people move to Seattle every week, most of which are working at Amazon or related companies. Not surprisingly, therefore, the population of the South Lake Union/Denny Triangle neighborhood has increased 169 percent from 2010 to 2018, compared to an overall city population increase of 23 percent.[6]

But the characteristics of who lives in these downtown neighborhoods are what are significant. Notably, in the South Lake Union/Denny Triangle where Amazon's headquarters is located, median household income increased from $30,000 in 2010 to $100,300 in 2018.[7] Citywide, median household income was $93,500 in 2018, up $33,000 since 2010, and software developer is now the most common occupation.[8] Educational attainment is up in neighborhoods near Amazon's headquarters, with a 12 percent increase since 2010 in residents with a bachelor's degree, while the percentage of African American residents in the Central District went from 64 percent in 1980 to less than 20 percent in 2016.[9] There are, of course, complexities to these changes in the racial makeup of the city. Seattle has been reported as slightly "less white" overall in 2017, though those changes are the result of more people identifying as more than one race and a higher percentage of people identifying as Asian.[10] While demographers are still examining the racial changes that have accompanied Amazon's location in central Seattle, it is clear that issues of housing affordability are driven not only by an influx of white residents but also by wealthier ones of several races.

Most importantly, with all of this demographic change has come dramatic shifts to the housing market. Described as a "crisis" by both the community members I talk to and the local media, housing prices in Seattle have risen 2.3 times faster than the national average recently, and between 2010 and 2017, rents increased the most of any city in the United States (64 percent).[11] Single family home prices rose 13.5 percent in 2017, and the median home price was more than $730,000 in 2017, compared with $350,000 in 2012.[12] Correspondingly, Seattle's rental market is also among the fastest rising in the United States, and much of the city's older, affordable housing rental and apartment stock is being purchased by investors to be redeveloped into new, taller, more expensive luxury condo buildings. This loss of affordable housing is happening at the same time that the homeless population in Seattle was reported to be between 6,000 and 8,000 individuals.

Unequivocally, since Amazon expanded into downtown Seattle, an affordable housing crisis, dramatic changes to local culture, and a rapid transformation of the built and natural environment have also occurred. The *Seattle Times* notes this phenomenon is increasingly called "Seattle-ization" to describe the rapid social and economic shifts that are accompanying the expansion of big tech in cities around the nation. The *New York Times* even cautioned HQ2 finalist cities of Seattle-ization, saying they may not want to become "too tech-y" like Seattle and warned that "the high housing costs and tech riches there have remade the city [Seattle] with startling speed."[13]

It cannot be overstated how much Seattle has changed since the arrival of Amazon downtown, and as someone who researches the city, I have witnessed and experienced these changes firsthand for over a decade. While Seattle's "green" reputation remains, the once gritty and artsy aspects of the city are being replaced with fancy and fast-paced images. While all cities undergo changes, and it is not reasonable to have nostalgia for a particular way of life that does not evolve over time, Seattle has experienced shifts in its population, housing stock, and local economy that have made the city far more exclusive and unaffordable. The stories of longtime Seattle residents being forced out of the community find their way into all kinds of local independent publications, social media groups, and community conversation. This is consistent with research that shows Seattle is the third most rapidly gentrifying city in the United States, as of

2019.[14] While many of the new highly paid technology workers get to enjoy Seattle's booming economy, exciting nightlife, expansive public transit, and Northwest culture, lower-income residents are increasingly shut out and displaced. In this context, it is important to explore two questions about the phenomenon of Seattle-ization. First, what allows this to happen, and how can Amazon change a city so much in such a short time period? Second, can it be resisted?

UNDERSTANDING AMAZON'S RISE: THE REAL ESTATE STATE IN THE SETTLER COLONIAL CITY

Literature in urban studies on the "neoliberal" or "entrepreneurial" city has been around for some time to help us describe the power of corporations over local governments. This literature notes that, under a shift to neoliberal governance in the 1970s and 80s that eroded the national welfare state and prioritized capital expansion, municipal governments became much more responsible for both economic development and social welfare. As far back as 1993, Kevin Cox identified the "new urban politics" that left cities to manage the "hypermobility of capital, and immobility of urban communities."[15] In this context, it was noted how municipal governments became captive to the capitalist hamster wheel of investment and disinvestment that forced cities to do whatever it takes to lure business (and a tax base) to their jurisdiction. Tim Hall and Phil Hubbard similarly noted, "There has been a reorientation of urban governance away from the local provision of welfare and services to a more outward-orientated stance designed to foster and encourage local growth and economic development."[16] Put simply, urban scholars have clearly demonstrated how and why business interests began to take precedence over those of the community—especially those people with less economic, political, and social capital.

How entrepreneurial cities try to attract capital to their jurisdictions has continued to be examined since this earliest research. For example, the selective incorporation of ecological and sustainability goals into local economic development agendas has been a means to attract and retain specific industries and specialized workers.[17] Scholars have also noted the use of "city branding" and marketing techniques to differentiate cities in a highly competitive landscape.[18] The adoption of "smart city" technologies (e.g., use of data to enhance urban life) has also been noted as a key strategy for attracting industry, especially new creative and technology companies. This has created "a diversion of public (welfare) resources to help lure in mobile global capital thereby creating social polarization."[19] All of these have been present in Seattle's evolution with Amazon over the past decade, as well as in many other cities throughout North America.

In this context, it is helpful to identify key mechanisms that allow corporations to transform urban spaces for their own benefit. That is, neoliberal urbanization is facilitated by the "real estate state" that caters to the interests of developers and private property owners over nearly all other aspects of urban life.[20] Also, contextualizing the real estate state as part of the settler colonial city helps us identify the underlying property relations that produce ongoing cycles of dispossession and displacement. By integrating these ideas into urban theory, we can more carefully understand differentiation, dispossession, and resistance in a place such as Seattle.

LOCAL GOVERNMENT AND THE REAL ESTATE STATE

Arguably, no concept has been more theorized, examined, and evaluated by social scientists than "the state," so I am deeply cautious about bringing a new formulation of the state to this analysis. Yet Samuel Stein's articulation of the "real estate state" is very helpful for bridging ideas regarding land and property—especially as they are related to displacement and gentrification—into conversation with neoliberal forms of local governance. Stein defines the real estate state in this way: "The price of land becomes a central economic determinant and a dominant political issue. . . . Government, particularly at the municipal level, becomes increasingly obsessed with raising property values and redistributing wealth upward through land and rents. . . . [W]e witness the rise of the *real estate state*, a political formulation in which real estate capital has inordinate influence over the shape of our cities, the parameters of our politics and the lives we lead."[21]

Stein's configuration of the real estate state clearly demonstrates that corporate and real estate interests have co-opted the state, and especially the local state, to prevent any meaningful regulation or intervention into private property relations. In a city such as Seattle, the forms of gentrification produced by developers, and supported by the real estate state, displace not only people of color but also other impoverished and unhoused people who are often subjected to laws and ordinances prohibiting loitering or panhandling.[22] A more detailed discussion of *how* the real estate state operates in the settler colonial city is not possible to provide here, but it includes things such as technocratic training of urban planners and public consultation processes that often, though not always, are dominated by the interests of the real estate state.

This helps us understand how Amazon leverages the real estate state to justify its political dominance because the higher-income jobs they create help increase property values, while the real estate state uses Amazon's influence over local government to prevent adjustments in property relations when people want change. Amazon is by no means the first company to take advantage of this; there is a long history of the power of capital and land development in the Seattle region from the earliest extractive industries to Boeing, Microsoft, and Vulcan. All of these companies helped to further entrench the private property relations, real estate investments, infrastructural developments, and cycles of housing dispossession now associated largely with Amazon. The developers driving gentrification and the real estate state that enables them are doing exactly what they are supposed to do—increasing the value of private property—as they both profit off of Amazon's role in driving "local economic development." But is this an artifact of the past half century of neoliberal urbanization, or can we link this to longer processes that have shaped the city?

PROPERTY RELATIONS IN THE SETTLER COLONIAL CITY

To fully understand how Amazon and the real estate state are able to transform a city such as Seattle for their own benefit, we must first understand property relations as part of an ongoing project of settler colonial urbanization.[23] This framework calls attention to land and property relations that underpin cycles of dispossession and displacement in cities. That is, colonization is not an event of the past on which cities are built but is very much a project of the present, in which cities are key sites in ongoing land theft and expendability of racialized bodies that constitute previous forms of colonization. Settler colonization shows that "through property, dispossession and settlement are continuously performed and creatively enacted."[24] In Seattle, we can trace this settler colonial logic from the removal of the Duwamish people, through segregation and redlining, to the rapid gentrification of low-income neighborhoods and communities of color today. A "settler class" of predominantly white residents with overwhelming control of both property and wealth is distinguished from communities of color and other marginalized groups who are subjected to a variety of policies and programs that sustain inequality. Specifically, Peter Kent-Stoll brings together the racial and colonial dimensions of gentrification by arguing that "Black populations are targeted by white settler logics as white spaces are produced through settler violence and worldviews, compounding both anti-Indigenous and anti-Black racism."[25] This allows us to see gentrification in the neoliberal city not as an artifact of the past half century but as part of much longer social and economic processes deeply connected to race and land.

Gentrification, urban redevelopment, and dispossession, then, are all important aspects of land theft in the colonial city and should be understood as more than an economic process.[26] They are also a project of violence rooted in land dispossession of many forms, including evictions, rent hikes, property tax increases, and redevelopment toward high-end housing. In contemporary settler colonial cities, actions of the settler class are "aimed primarily at exploiting the wealth and resources of a colonized space . . . driven by a desire to possess and settle land in perpetuity, in addition to benefiting from new opportunities to accumulate."[27] The real estate state further allows us to understand "how the settler colonial politics of dispossession fused with the legal architecture of the settler state to produce a racialized system of property ownership, designed not only to facilitate the theft of Indigenous land, and the establishment of settler policy, but also enact a massive transfer of wealth."[28] To be clear, the racial dynamics of who causes gentrification in Seattle are complex: people of color are both gentrifiers and subject to gentrification. Yet there is a distinct settler class in the city, all of whose "claims to land and citizenship are entirely dependent on an ongoing structure of Indigenous land dispossession," in which non-Indigenous settlers have a "sense of connection, if not sense of entitlement, to their new home."[29] When it comes to Amazon, we must critically evaluate these private property relations that undergird the settler colonial city and continue a cycle of racialized dispossession and displacement.

In this context, housing is not seen as a human right but rather a massive profit generator for the settler class, a full-blown commodity that requires a state apparatus to explicitly and unwaveringly protect the settler class and maintain deeply entrenched socially unjust private property relations. As a result, characteristics of the real estate state in the settler colonial city include ongoing cycles of violent dispossession and displacement; active production of segregated and exclusionary urban spaces; commodification of Indigenous land, especially for housing; commodification of "diversity" and "sustainability" to raise property values; policies that protect property values over people; and erasure of non-settler cultures, communities, and alternative economic relations.

While these aspects of settler colonial urbanization that continue to shape our cities today are deeply problematic, they are never fully complete. Social justice groups in Seattle and beyond are actively working to resist both the real estate state and settler colonial urbanization, and they provide key insights into how to forge a path of resistance.

HOUSING JUSTICE ACTIVISM IN SEATTLE

> Amazon, corporate developers, and the other super-wealthy corporations in Seattle, who are raking in tens of billions in profits annually, must be taxed to fund this new [affordable] housing—not working families and small- and medium-sized businesses.
>
> —KSHAMA SAWANT,
> Socialist city councilmember

The Amazon "head tax" controversy can help us understand how resistance is formed in relation to the real estate state and settler colonial urbanization in Seattle. In May 2018, after much work and organizing by a number of people and organizations, the Seattle City Council unanimously adopted an employee hours tax (head tax) that would levee "a $275 per employee tax on Seattle businesses making more than $20 million a year" to create a $40 million per year fund to be used for building affordable housing and providing homeless support services in the city.[30] A direct reaction to Amazon's role in the affordable housing crisis, the head tax was celebrated as a significant game changing action across the United States.

Yet, before citizens could vote in a referendum to approve the tax, Amazon sprang into action to intimidate and bully city councilmembers to repeal the tax. In interviews, people reported to me that Amazon threatened not to move into Amazon's newest headquarters building, thereby pulling thousands of jobs, if the tax was implemented. Interviewees also told me that hundreds of millions of dollars were spent to advertise and campaign against the tax. Ultimately, before the issue could even go to the voters, the mayor supported a repeal of the tax. The city council approved the repeal, with only two councilmembers voting against (Kshama Sawant and Teresa Mosqueda). Amazon, big business, and the real estate state had achieved a stunning and disturbing victory.

Despite the hegemonic presence of Amazon in the settler colonial city and the company's skilled engagement with the real estate state, robust and significant housing justice activist groups have emerged. Unlike more mainstream approaches to affordable housing that simply seek to increase the supply of housing, these Seattle activists demand transformative solutions to the housing and homeless crisis as they reconceptualize property relations in the city, including housing as a human right. These efforts engage a variety of tactics and strategies, including two that I will discuss here: 1) intentional social movement building to challenge the real estate state, and 2) political disruption that engages the practice of agitation to recast who "talks" and who "listens" in the settler colonial city. While I make a distinction between these aspects of the housing justice movement to specify how this activism takes place, social movement building and political disruption do occur simultaneously in Seattle.

What is reported below comes from my personal and sustained engagement with housing justice activists in Seattle. This includes interviews and close observation of housing justice activities (2015–present). The information presented here is also based upon a longer-term research project on local climate governance in Seattle that started in 2007, which has allowed me to see the significant social and physical changes in the city since Amazon located its headquarters there the same year. My research and the ideas here reflect a commitment to prioritizing the knowledge, experience, and visions of marginalized communities.

MOVEMENT BUILDING

The chambers of city hall have become a significant and productive site for movement building, led in large part by Councilmember

Kshama Sawant and the Socialist Alternative Party. For example, in August and September 2016, a group calling itself Block the Bunker overfilled the council chambers on several occasions to demand that $149 million earmarked for a new police substation in north Seattle be redirected immediately and completely for affordable housing in the city. Protestors remade the public discourse from one about routine actions for approving funding for a police station to one that considered the intersections of racism, police brutality, homelessness, and an affordable housing crisis in the city. By framing the proposed police station as a "bunker," they were able to contest the normalization of a police state and racialized violence by police in the midst of very little action on the affordable housing and homeless crisis in the city.

In another movement building effort, more than one hundred people turned out for the launch of Affordable Seattle—a coalition of various social justice groups such as Black Lives Matter and Socialist Alternative—on July 29, 2017. The group's platform consists of three key pillars: building affordable housing by taxing business, pushing for rent control, and making landlords pay the moving costs of residents who are forced out via economic eviction when rents rise above their ability to pay. Speakers at the launch event included mayoral candidate Nikkita Oliver (People's Party) and city council candidate Jon Grant (Democratic Socialist Party). The event was an important catalyst to bring together people from various social justice organizations in the spirit of exchange, momentum building, inspiration, and community building. The group has helped popularize the phrases "we need rent control" and "tax Amazon now" throughout the city of Seattle.

Similarly, the group Vanishing Seattle acts as a coordinating hub for all those who despise the ongoing transition of Seattle as it "documents the displaced & disappearing institutions, small businesses, & cultures of Seattle."[31] Their activities also include meetups, art performances/installations, storytelling, short films, speaking engagements, and a lively social media presence. In all of these efforts, the group remakes the narrative of "economic development" into one of loss, nostalgia, and mourning. In a personal engagement with the group, I learned from the group's Facebook page that the Two Bells Tavern in downtown Seattle became one of the latest victims of the city's rapid gentrification as it closed its doors. This was a place that

I regularly visited during every research trip I took to Seattle to both relax at the end of a long work day and to talk with long-term residents about how they felt about the changes in the city. Not only was the establishment a beloved seventy-four-year-old bar and restaurant known for cheap beer and food, it was an important gathering place for non-techy Seattleites who still lived downtown to spend time together. It is in this context that groups such as Affordable Seattle and Vanishing Seattle provide a way for Seattle residents to better understand the connections between a changing city, the affordable housing crisis, and the technology sector boom in new and substantial ways.

POLITICAL DISRUPTION

On any given day, there are hundreds of new real estate development projects occurring throughout Seattle. As noted, most of these projects include new luxury residences, office units, and commercial space, and there are only a small number of developers explicitly building "workforce housing."[32] As such, there are hundreds of "Proposed Land Use Action" signs that are posted in front of new multimillion-dollar residential and commercial developments throughout the city, which are required by the local planning office to note when a building is being redeveloped by a real estate company. These signs describe the proposed project and provide contact information for the developer and relevant city departments.

It is nearly impossible to walk a single block downtown or in the adjacent neighborhoods without passing at least one of these signs. But what is also clearly noticeable is that all of them are covered in graffiti and stickers mocking the developments and lamenting the "takeover" of old Seattle by big corporate money. The defacing of these signs is so rapid and pervasive that the city's Department of Construction and Inspections acknowledges they are almost of no use and has repeatedly redesigned them to try and make them more functional. For example, long-time residents of Capitol Hill and other centrally located neighborhoods are fed up with the rapid transformation of their neighborhood. They protest in many ways, including posting stickers on the yellow land use action signs that say "STOP THE DESTRUCTION, SAVE OLD SEATTLE, BOYCOTT NEW BUILDINGS. Take back this city from our very, very corrupt elected officials and those greedy real estate developers, and greedy property owners." These are acts of defiance that are intended to render visible the frustration with developers and with the real estate state.

Local activists have also become quite effective in utilizing the required public comment period at city council meetings. During the Block the Bunker protests at city hall, activists enacted intentional forms of resistance, conducting themselves in a manner that provides clear disruption of formal government procedures: waving large red signs, wearing coordinated T-shirts, clapping, booing, wearing masking tape over their own mouths, dressing up as a cop with a pig nose, crying, and cussing—generally causing disruption to call attention to crisis. For example, at the August 15, 2016, Seattle City Council meeting, the following statements (often spoke by women and people of color) were made during public comment: "make developers pay their fair share," "we need jobs, not jails," "divest from white supremacy," "bull shit bunker," "we need to give the land back to the Indigenous," and "stop the war on Black lives."[33] Attendees to the city council meeting also asked their elected leaders for eye contact when they were speaking to them, to remove a racist heckler, and to let in a number of people who were blocked from entering the council chambers. Reports of the event suggested that some elected officials in Seattle were increasingly agitated with the significant and direct displays of resistance by activists. It should be noted that Councilmember Kshama Sawant and the Socialist Alternative Party are central to these organizing efforts.[34] Councilmember Sawant has been instrumental in shifting the political landscape in Seattle to see political disruption as necessary to resist the real estate state in the settler colonial city.

Examples such as these demonstrate the forms of intentional social movement building that are working to resist the real estate state and create discursive shifts that recast who "talks and who listens" in the settler colonial city. Using the tactics described here, protesters in Seattle have mobilized a mass movement of residents demanding affordable housing. Currently, a massive "Tax Amazon" campaign is being built in the city, which shows increasing promise for creating change.

RESISTING SEATTLE-IZATION BEYOND CASCADIA

The most evident example of how Seattle-ization extends beyond Cascadia requires returning to Amazon's HQ2 search. While status quo capitalists and elected officials celebrated the news that Long Island City was one of the winners of HQ2, local residents worked to resist the occupation of their community by Amazon. Championed by Representative Alexandria Ocasio-Cortez, activists organized themselves to make sure Amazon knew they were not welcome, primarily because of their role in spurring gentrification and displacement. For example, when I visited the neighborhood shortly after the announcement, I saw all over the community spray-painted signs saying "Amaz**no**" and an inverted sign of the Amazon smile logo to show a frown. Activists also held protests and rallies in which they called out politicians for doing a "backroom deal" for HQ2, giving away incentives the community had not agreed to give. And, not surprisingly, the Queens neighborhoods where HQ2 would be built are ones ripe for gentrification by the settler colonial corporation and the real estate state: lower income and predominantly filled with residents of color.

Remarkably, Amazon announced they would pull HQ2 out of Long Island City, as Bezos boasted that the city would regret the lost jobs and economic development. It is unclear exactly what prompted him to leave, but it was a victory for the residents of Long Island City against the real estate state and a new wave of settler colonial urbanization. Activists reported being both terrified of becoming a victim of Seattle-ization and also inspired by housing justice activists actively resisting Amazon in Seattle.

CONCLUSIONS

This chapter was written before the COVID-19 pandemic caused new waves of economic crisis and displacement in Seattle. It was also written before the 2020 racial justice protests in Seattle, at which activists demanded the reformation of policing, housing, and other racist institutions. It remains to be seen how these events might reshape the city, but despite those uncertain futures, the ideas and analysis presented here are intended to show why the real estate state and its close connections with settler colonial urbanization require careful and sustained attention in the age of Amazon. While very evident in Seattle, these issues are not confined to the big "tech cities" of Seattle, Silicon Valley, Austin, or New York. Instead, they extend to communities across North America where forms of settler colonial urbanization have dominated urban communities for centuries and where a deeply entrenched real estate state props up these violent processes, ensuring settler populations continue to benefit from the expendability and dispossession of others.

Despite these difficult realities, I have also considered the ways that justice movements in Seattle have influence in and beyond Cascadia, especially in Long Island City, New York, where activists helped force Amazon to rescind its decision to locate HQ2 in their neighborhood. So, while we can see that technology, creative, and information industries are the latest articulation of the settler colonial city and the real estate state, we know that this project is never fully complete and people bravely resist it all the time.

I want to close with some concrete thoughts about what we learn from these activist communities to resist the real estate state and settler colonial urbanization. First, we must all adamantly reject that gentrification is "just the way it is" and demand that our elected officials and decision-makers stop insisting we can "manage" the problems with the tools offered by the real estate state. We must reject private land markets, especially those for housing, by both creating new forms of community (not-for-profit) landholdings and passing strong private property regulations that reduce the opportunity and incentive for developers to displace vulnerable residents. This can include community land trusts, rent control, hefty corporate taxes that are used to build massive amounts of public, affordable housing, or prohibitions on developers that flood the housing market with luxury homes.

Second, I strongly believe that the feelings of loss of community by those subjected to the violence of settler urbanization are *not* forms of NIMBYism ("not in my backyard") or fear of change, as real estate developers and elected officials often say. Rather, as my analysis shows, they are genuine feelings tied to racial violence, land dispossession, cultural commodification, and social erasure enacted by the real estate state. We must continue to render visible these feelings and experiences, even in the face of massive opposition by big corporations and the real estate state.

Finally, we need genuine and meaningful forms of redress that remake the private property regulations of the settler colonial city and offer a chance for healing among those harmed by it, starting with Indigenous communities both living in our cities and displaced out of them. This means there must be consent in how land is transformed among a much broader constituency. This will require that we listen to vulnerable communities to help them enact a range of reforms, such as land reform, state reform, and economic reform, as we develop new forms of radical democracy and demand racial justice to finally end the real estate state.

NOTES

The chapter's epigraph is from Dennis Green, "'We Shouldn't Be Inviting Bullies to Our Neighborhood': Alexandria Ocasio-Cortez Defends Her Criticism of Amazon HQ2 in Surprise Queens Appearance," Business Insider, March 20, 2019, https://www.businessinsider.com/alexandria-ocasio-cortez-defends-amazon-hq2-criticism-2019-3. Gene Balk epigraph is from the January 2, 2019, *Seattle Times*. Kshama Sawant epigraph is from "Tax Amazon," Seattle City Council, City of Seattle, April 17, 2020, www.seattle.gov/council/meet-the-council/kshama-sawant/tax-amazon.

1 Amazon, "Amazon HQ2 RFP," September 2017, https://images-na.ssl-images-amazon.com/images/G/01/Anything/test/images/usa/RFP_3._V516043504_.pdf.

2 Jason Del Rey, "Amazon Received More Than 200 Proposals from Places That Want to Host Its New Headquarters," Vox, October 23, 2017, www.vox.com/2017/10/23/16521500/amazon-hq2-new-second-headquarters-rfp-proposals-bids; and Alison Griswold, "A Nearly Complete List of the 238 Places That Bid for Amazon's Next Headquarters," Quartz, November 4, 2017, https://qz.com/1119945/a-nearly-complete-list-of-the-238-places-that-bid-for-amazons-next-headquarters/.

3 Samuel Stein, *Capital City: Gentrification and the Real Estate State* (London: Verso Books, 2019).

4 Jennifer L. Rice et al., "Contradictions of the Climate-Friendly City: New Perspectives on Eco-gentrification and Housing Justice," *International Journal of Urban and Regional Research* 44, no. 1 (2020): 145–65.

5 Rice et al., "Contradictions of the Climate-Friendly City."

6 Gene Balk, "5 of the Seattle Area's Most Changed Neighborhoods: We Crunched the Data On Population, Income, Jobs," *Seattle Times*, January 21, 2020.

7 Balk, "5 of the Seattle Area's Most Changed Neighborhoods."

8 Gene Balk, "Median Income in Seattle Skyrockets to $93,500," *The Columbian*, September 26, 2019.

9 Rice et al., "Contradictions"; Tyrone Beason, "Central District's Shrinking Black Community Wonders What's Next," *Seattle Times*, May 28, 2016.

10 Gene Balk, "Seattle Is Less White Than It Has Ever Been in Modern History, New Census Data Show," *Seattle Times*, September 14, 2017.

11 Mike Rosenberg, "Seattle No. 1 in Home-Price Growth Again; Starter Homes Require Half of Income," *Seattle Times*, June 27, 2017; Mike Rosenberg, "Rent Hikes Slow around the Country, but Surge Again in Seattle. What's Going On?" *Seattle Times*, June 27, 2017.

12 Chris Morris, "This Is the Hottest Real Estate Market in the Country—by Far," *Fortune*, September 27, 2017.

13 Emily Badger, "Happy New Year! May Your City Never Become San Francisco, New York, or Seattle," *New York Times*, December 26, 2018.

14 Gene Balk, "Seattle Is the Third Most Gentrifying U.S. City—but That Might Not Be as Bad as You Think, Study Finds," *Seattle Times*, July 24, 2019.

15 Kevin R. Cox, "The Local and the Global in the New Urban Politics: A Critical View," *Environment and Planning D: Society and Space* 11, no. 4 (1993): 433–48; see also David Harvey, "From Managerialism to Entrepreneurialism: The Transformation in Urban Governance in Late Capitalism," *Geografiska Annaler: Series B, Human Geography* 71, no. 1 (1989): 3–17.

16 Tim Hall and Phil Hubbard, "The Entrepreneurial City: New Urban Politics, New Urban Geographies?" *Progress in Human Geography* 20, no. 2 (1996): 153–74.

17 Anthony M. Levenda and Eliot Tretter, "The Environmentalization of Urban Entrepreneurialism: From Technopolis to Start-up City," *Environment and Planning A: Economy and Space* 52, no. 3 (2019): 490–509; and Aidan While, Andrew E. G. Jonas, and David Gibbs, "The Environment and the Entrepreneurial City: Searching for the Urban 'Sustainability Fix'in Manchester and Leeds," *International Journal of Urban and Regional Research* 28, no. 3 (2004): 549–69.

18 Mihalis Kavaratzis, "City Marketing: The Past, the Present and Some Unresolved Issues," *Geography Compass* 1, no. 3 (2007): 695–712.

19 Robert G. Hollands, "Will the Real Smart City Please Stand Up? Intelligent, Progressive or Entrepreneurial?," *City* 12, no. 3 (2008): 314.

20 Stein, *Capital City*.

21 Stein, *Capital City*, 5, emphasis in original.

22 Jean-Paul D. Addie and James C. Fraser, "After Gentrification: Social Mix, Settler Colonialism, and Cruel Optimism in the Transformation of Neighbourhood Space," *Antipode* 51, no. 5 (2019): 1369–94.

23 David Hugill, "What Is a Settler-Colonial City?," *Geography Compass* 11, no. 5 (2017); and Libby Porter and Oren Yiftachel, "Urbanizing Settler-Colonial Studies: Introduction to the Special Issue," *Settler Colonial Studies* 9, no. 2 (2019): 177–86.

24 Naama Blatman-Thomas and Libby Porter, "Placing Property: Theorizing the Urban from Settler Colonial Cities," *International Journal of Urban and Regional Research* 43, no. 1 (2019): 30–45, citations omitted.

25 Peter Kent-Stoll, "The Racial and Colonial Dimensions of Gentrification," *Sociology Compass* 14, no. 12 (2020): 1–17, https://doi.org/10.1111/soc4.12838.

26 Addie and Fraser, "After Gentrification."

27 Hugill, "Settler-Colonial City," 2, 4.

28 Heather Dorries, David Hugill, and Julie Tomiak, "Racial Capitalism and the Production of Settler Colonial Cities," *Geoforum* (2019): 5, https://doi.org/10.1016/j.geoforum.2019.07.016.

29 Janice Barry and Julian Agyeman, "On Belonging and Becoming in the Settler-Colonial City: Co-produced Futurities, Placemaking, and Urban Planning in the United States," *Journal of Race, Ethnicity and the City* 1, no. 1–2 (2020): 22–41, https://doi.org/10.1080/26884674.2020.1793703.

30 Alana Semuels, "How Amazon Helped Kill a Seattle Tax on Business: A Levy on Big Companies to Fund Affordable Housing Awakened the Ire of Corporations," *The Atlantic*, June 13, 2018. It was reported to me in an interview that this amounted to only 2 percent of Seattle businesses that would be subject to the head tax.

31 "About," Vanishing Seattle, accessed January 27, 2021, www.vanishingseattle.org/about.

32 Rosenberg, "Rent Hikes."

33 Dyer Oxley, "Tempers Flare over Police Precinct at Seattle Council Meeting," MYNorthwest, August 16, 2016, http://mynorthwest.com/370464/council-halts-north-precinct-meeting/.

34 Jennifer L. Rice, "'The Everyday Choices We Make Matter': Urban Climate Politics and the Postpolitics of Responsibility and Action," in *Towards a Cultural Politics of Climate Change: Devices, Desires and Dissent*, ed. Harriet Bulkeley, Matthew Paterson, and Johannes Stripple (Cambridge: Cambridge University Press, 2016), 110–26.

CHAPTER 10

CONFLICTING SUSTAINABILITIES AND THE LIMITS OF LOCALIZED GREEN GOVERNANCE

CORINA MCKENDRY

THE MAJOR CITIES OF CASCADIA ARE ALL FACING A CRISIS of affordability.[1] For a variety of reasons, Seattle, Portland, and Vancouver have seen housing prices and rents skyrocket over the past decades. The growing income inequality that has accompanied the expansion of the knowledge and technology sectors has increased housing demand from highly skilled and highly paid residents.[2] International capital investment in the Cascadian housing markets has driven prices up further.[3] In addition, these cities' famous green amenities and environmental initiatives arguably contribute to increased housing costs by drawing the middle class and well-to-do back from the suburbs to amenity-rich, climate-resilient green cities.[4]

The lack of social justice represented by expensive housing and its accompanying gentrification and displacement is one of the deep challenges of urban Cascadia (see chapter 9). It raises questions of whom the green city is for as well as what can be done to ensure that all communities, including those who have benefited least from the riches of the knowledge economy, secure a place in a greener city. It also raises questions as to local governments' abilities to create just, sustainable cities.

Though the pressing injustice of a lack of affordable housing is a regional challenge, this chapter helps unpack this issue through examining development controversies in three neighborhoods of Vancouver, British Columbia, in the mid-2010s. These cases illustrate ways that a municipal government with a history of robust (though not uncontroversial) public participation in city planning addresses conflicts around the form and process of green urbanization. Of notable importance is the tension between the desire of communities to maintain the connections they feel to their neighborhoods through limiting changes in demographics and the built environment and the city's focus on increasing affordable housing and reducing carbon emissions through rapid densification.

The difficulty Vancouver faces in resolving this tension lies only partly in the market-based orientation and green "sustainability fix" that have shaped and limited many urban greening processes.[5] Equally significant are deep disagreements among city residents as to what a just, sustainable city is and how it should be achieved. As climate scientists assert the need for accelerating decarbonization in the face of climate change, and as the crisis of housing affordability demands substantive responses, Vancouver illustrates how efforts to achieve such transformations can run against people's sense of time, place, and lived experience, with challenging implications for the role of the local state in furthering socially just, green urban governance.

DENSITY AND THE SUSTAINABLE CITY

Vancouver, BC, has been heralded as the world's most livable city and one of the world's most beautiful cities, and city leaders have the goal of making it the world's greenest city.[6] It is famous for its history of urban planning, wherein high-density development has been combined with the provision of high-quality environmental amenities, including extensive public access to the waterfront; plentiful parks; and pedestrian, bike, and transit-friendly infrastructure.[7] The city is also incredibly expensive, and for a number of years, Vancouver has been ranked one of the least affordable cities in the world.[8]

Vision Vancouver (henceforth Vision), the municipal political party that controlled the Vancouver mayor's office and the city council from 2008 to 2018, adopted an ambitious environmental and social agenda that included increasing affordable housing, ending street homelessness, dramatically decreasing carbon emissions from buildings and transportation, and eliminating waste sent to landfills. One of the principal strategies that the mayor, Gregor Robertson, and the Vision-controlled council saw as key to many of these goals was expanding housing supply. In the tightly constrained boundaries of the Vancouver city limits, this meant increasing density.

Throughout Vision's ten years controlling city hall, densification and development became major flashpoints of political controversy. The debates could be summarized as follows: city councilors and planners saw densification as vital to the future of the city. For them, it was the only way to reduce carbon emissions through more energy efficient buildings and district heating systems, enable enhanced green transportation, provide housing for the city's expected 150,000 new residents over the next three decades, and increase access to affordable housing for the poor and the middle class. On the other side of the debate were vocal community members and opposition politicians who saw the claim that density was key to city sustainability and affordability as a façade that officials were using to push through developers' agenda of excessive density in inappropriate places and against community wishes. City officials, in return, saw resistance to new development as, at best, a misunderstanding of the scale of change that needed to occur to achieve Vancouver's social and environmental goals or, at worst, a thinly veiled effort by wealthy homeowners to keep renters and lower-income families out of their neighborhoods in the name of community preservation. These debates are not unique to Vancouver. In the wider literature on green urbanism there are ongoing debates about if and when density can achieve social and environmental goals or if a focus on urban form ignores more important issues of housing as a commodity, jobs and living wages, and so on. Yet the particulars of these three cases illustrate the on-the-ground challenges of synthesizing the need for community control over the planning process with movement toward greener, more inclusive urban development.

PLANNING THE GREEN CITY

Broadly, there is evidence that cities are most enthusiastic about environmental initiatives that have clear economic benefits. They are much less interested in local environmental regulations that could discourage growth or that entail substantial costs.[9] Yet this generalization hides notable differences between city leaders' willingness to push the limits of neoliberal environmentalism.

In Vancouver, though undoubtedly maintaining a commitment to economic growth, leaders have been decidedly more willing than those in most other major green cities to adopt stringent mandates on developers to further the city's environmental and housing agendas.[10] This has included the passage of the strictest green building standards in North America and recycling requirements for demolished buildings, negotiating robust community amenity contributions and the provision of affordable housing in exchange for building permits and upzoning, and more.[11] Facing skyrocketing property values, little provincial funding for affordable housing, and minimal taxing authority, Vision officials saw their control over land use and new development as one of the only tools they had to meet the growing city's needs.

Density accompanied by amenities as a way to achieve social and environmental goals did not begin with Vision. It began decades earlier, particularly with the redevelopment of the acres of waterfront land that were sold to the development corporation Concord Pacific after the World's Fair held in Vancouver in 1986 (known as Expo 86). In the years that followed, development pressure started to be felt in established residential neighborhoods across the city. In response, in 1995, a process of community-based neighborhood planning was codified through CityPlan. At the heart of CityPlan was the idea that each community should be able to decide how it wanted to develop in order to increase density in a way that met the needs of the city's growing population while also enhancing neighborhood sustainability, livability, and identity. In theory, the development plans that emerge from this process would be quite democratic, with extensive participation and input from all types of residents of the neighborhood leading to a widely accepted plan that reflects the vision and priorities of the community.[12]

CityPlan's slow progression toward a denser urban form became articulated even more explicitly with the city council's approval of the EcoDensity Charter in 2008. EcoDensity was touted as "an unprecedented planning effort to use density, design, and land use as catalysts for an environmentally sustainable, affordable and livable Vancouver."[13] More specifically, the goal of EcoDensity was to increase density along transit routes in neighborhoods consisting mainly of single-family homes. EcoDensity was incredibly polarizing and engendered two years of public debate.[14] Proponents claimed EcoDensity was merely a rearticulation of the goal of CityPlan to achieve sustainability and livability through walkable neighborhoods and mixed-use development.[15] Opponents countered that the lack of public input into EcoDensity was contrary to the neighborhood planning process of CityPlan and that it would mean a top-down imposition of unwanted density in neighborhoods throughout the city.

The term *EcoDensity* was dropped with the election of Robertson and the Vision-controlled city council shortly after the EcoDensity Charter's passage. However, in interviews in 2014, Vision officials pointed to the politically contentious legacy of EcoDensity as one of the sources of the ongoing struggle around densification. One Vision councilor described the impact of EcoDensity on Vancouver planning politics in this way:

> What [the former mayor] did [when he announced EcoDensity with little public input] was he freaked people out. He terrified people by saying: "we are going to densify every neighborhood"—which we are! But he did it by announcing it as a done thing. . . . If instead . . . [he had said], "Ten years ago we did the CityPlan . . . it was fantastic, people! . . . [But] in the last ten years, here's what's happened. People are moving here faster than we'd anticipated. Climate change has become a thing. The affordability of our land has gone through the roof. . . . Let's add those layers onto the maps . . . and see what happens. . . . Let's talk about what form has to follow that function. Of absorbing these people, of decreasing our footprint, of not sprawling into the agricultural land, or the ocean, or the US, or the mountains." . . . But it didn't happen that way. . . . So that's one of the challenges; it's that EcoDensity kind of made [density] an even dirtier word than it might have been if we'd let people come to the conclusion that densifying has to happen.[16]

It is impossible to know whether this councilor was correct and if a more participatory process of creating EcoDensity would have blunted some of the popular resistance to the idea of creating greater density outside the downtown core. What certainly is clear is that as the Vision council and Mayor Robertson began promoting densification as necessary to achieving low-carbon and housing affordability goals, strong, though not always consistent, opposition emerged. This was seen most notably in the spaces for contestation available through the community planning process.

CONTESTATION IN VANCOUVER'S COMMUNITY PLANNING PROCESSES

The different ways the tensions between affordability, environmental sustainability, and people's connection to place unfolded and intersected with Vision's development agenda during this era can be seen in the controversies surrounding the community development plans for the neighborhoods of Grandview-Woodland, Marpole, and the Downtown Eastside. In some ways, the controversies around these three plans were quite different. In Grandview-Woodland, protests focused on the impact of condominium towers on the vibrant, mixed-income neighborhood. In Marpole, a group of vocal homeowners objected to new townhouses and apartments close to their single-family homes. In the Downtown Eastside, concerns focused on gentrification and on whether the community development plan would prevent or contribute to displacement of the area's low-income residents.

Yet the controversies over the three plans also shared something in common. In each case, opponents articulated an understanding of connection to their neighborhoods, of what would make their community sustainable, and of how they wanted it to change (or, more often, not change) that was different from the vision offered by the city. Responding to residents' engagement and protests, city officials adjusted the development plans. Whether this was merely to placate discontent or a genuine effort to better balance residents' concerns with the local government's citywide priorities of increasing the housing supply is not clear. What is clear from the three planning processes, however, is the difficulty of defining, let alone creating processes to achieve, a just and sustainable city. The deep disagreements among city residents as to how to realize the widely supported goals of an affordable, environmentally sustainable city are as key to understanding the challenges facing environmental justice in Cascadia's increasingly unaffordable cities as the more common, and also important, critiques of the contradictions within the cities' planning processes (see chapters 8 and 9).

PARTICIPATION, POWER, AND APPROPRIATE DENSITY

Grandview-Woodland is most known for its vibrant small business district along Commercial Drive. It is also a diverse neighborhood with immigrants from numerous countries and a substantial stock of aging, relatively affordable apartments. In 2013, there was a six-month process to create a development plan for the area. The community planning process itself, a number of local activists told me, felt like a good one, with widespread participation and input. What Grandview-Woodland residents thought was going to emerge was a plan that prioritized maintaining the diversity and small-scale feel of the neighborhood while allowing modest density increases to facilitate new affordable housing.[17]

Community members were therefore furious when what was presented to the city council by the planning department was not what they had envisioned. Though the plan did call for preserving the low-income rental properties and the feel of Commercial Drive, it also included a thirty-six-story condominium tower near the neighborhood's already busy light-rail stop as well as several other shorter towers throughout the neighborhood. The towers were way beyond anything that had been discussed, let alone supported by residents, during the community planning process.[18] For activists, the blame for the towers lay squarely with the mayor. According to one, "There was a beautiful planning process. . . . But something happened between the planner who was delivering that really well designed and effective and genuine process and receiving the content [from the community] and then the general manager of planning . . . who got a directive from the mayor's office to facilitate a thirty-six-story tower development at the corner of the neighborhood, no matter what. That's where you've got to . . . call the bullshit."[19]

Call it people did. In response to intense protest, the city withdrew the draft community plan. It established a citizens' assembly that was representative of the demographics of the neighborhood to draft a new plan. The citizens' assembly spent two additional years on the plan, which was adopted by the council in 2016. In the final plan, the thirty-six-story tower was gone, but the plan still contained more density than some residents wanted. And despite the role of the citizens' assembly, some activists argued that the revised plan was still not representative of the wishes of the community, was "bludgeoned through City Council by the developer-financed Vision Vancouver majority," and would lead to increasing property costs, and therefore displacement, in the neighborhood.[20]

The Marpole community plan also led to local protest. As in Grandview-Woodland, opposition focused on how the plan would change the overall feel of the neighborhood, though in this case, opposition was led by homeowners concerned about new rental and multifamily properties rather than high-end towers. At the far south edge of the city, much of Marpole consists of single-family, detached homes, the type of housing that led the exorbitant increase in home prices that Vancouver started experiencing in the early 2000s.[21] Yet it also has a lower median income than the Vancouver average and a substantial number of renters and recent immigrants.[22]

In Marpole, the community planning process was well underway when vocal opposition emerged. Concerns began to be voiced "after homeowners realized the city was proposing to upzone vast areas of single-family homes to allow for townhouses, duplexes and other higher-density housing. 'We feel like we got dropped on our heads,' [a local organizer] said. . . . 'People are feeling like they're under siege.'"[23] In response to protests, the city withdrew proposed developments in the area of Marpole that was of particular concern to the homeowners group, though the final community plan did include increased density along major arteries.

In both Marpole and Grandview-Woodland, vocal residents felt that the planning process did not lead to development plans that reflected the vision the communities had for their neighborhoods. Rather, activists felt that in the face of a city council and mayor determined to increase density, local concerns about how new developments would fit in with the existing communities were dismissed. The city reducing the density of the initial proposals in response to protest did not convince opponents that planners and elected officials cared about the residents' vision for the neighborhood. Rather, it was seen as further proof that the initial planning process had failed to consider their input.[24]

Unsurprisingly, city officials offered a far different interpretation of the protests and their outcomes. In discussing the emergence of opposition to the Marpole plan, for example, one Vision city councilor expressed her frustration with the protesters:

> I met with [a leader of the protesting community group] and I said, "Tell me about what your issues are? Planning processes have been going on for two years; have you been involved with that? Hard for us to throw out two years of your neighbors' input because you're now aware of it. . . ." He says, "Well, there's no way my neighbors came up with this. . . . Nobody lives in anything other than a single-family home." And I was like, "How many renters are in your group?" And [he] said, "Well, I guess none, nobody rents in Marpole." And I was like . . . "53 percent of residents rent in Marpole," and he said, "That is just impossible" and "You've falsified the numbers." Unbelievable! And this is a guy all over the media talking on behalf of his community . . . and he actually believed no renters lived there! . . . In fact, the majority of Vancouver rents and most City Hall meetings are *not* dominated by renters.[25]

For this city councilor, a longtime housing justice and environmental activist, the disproportionate power of homeowners in Vancouver politics was limiting the construction of needed affordable housing in the city.

DISPLACEMENT AND PLANNING SUSTAINABLE COMMUNITIES

Though a very different kind of plan and neighborhood, the Downtown Eastside community plan also shows the challenge of defining—and achieving—socially sustainable development. It also illustrates the difficulties local governments confront navigating affordable housing goals in the face of property speculation and severe funding limitations.

The site of much of the city's single-room occupancy housing, the Downtown Eastside has long been home to many of Vancouver's poorest residents. Effective community organizing along with the severity of the area's social problems had long limited gentrification in the Downtown Eastside, but by the early 2010s, rising housing costs and the neighborhood's central location were leading to growing housing pressure.[26] A development plan was clearly necessary, but the diversity of the area, which includes not only the low-income Downtown Eastside but also tourism-focused Gastown and historic

Chinatown, meant that participants in the planning process had "a lot of disagreement on the future [of the] neighborhood."[27]

It took four years of work for the Downtown Eastside plan to be written and adopted.[28] The plan held that new development in the area where most of the services for low-income residents are concentrated must be 60 percent social housing and 40 percent market-rate rentals, envisioning 4,400 new units of social housing. Fees from developments on the edge of Chinatown would be used to pay for heritage revitalization, a priority for residents of that part of the neighborhood. New development in Gastown would continue to prioritize market-rate housing, with developer fees helping to fund affordable units elsewhere in the area. Planners saw the result as achieving an unprecedented balance between the differing community visions for the area, the difficulty of paying for low-income housing, and Vancouver's larger sustainability goals:

> Not everyone gets everything they want, but everyone gets something. And in the end, hopefully the balance comes out of the mix and results in a sustainable neighborhood. . . . I think it's incredibly aggressive, but it's doable. But it's barely doable. It relies on playing in the market in ways we've never had in the past, . . . throw[ing] everything we have at [developers]. . . . We have some pretty huge larger objectives: affordability, urban sustainability, energy, climate change, rising sea level. And all of those we have to try to bring down to the local level and make them make sense.[29]

Yet what planners and some community groups saw as a bold effort to facilitate social and environmental sustainability in the area, others saw as exacerbating Vancouver's problems with growth, development, and displacement. Some longtime housing advocates in the Downtown Eastside publicly supported the plan. They argued that considering the intensity of gentrification pressure and real estate speculation in the area, as well as the dearth of funding for social housing from higher levels of government, the plan was as good as could be hoped for and would help prevent the displacement of the neighborhood's low-income residents.[30]

Others disagreed. They saw the protections for low-income renters as inadequate for preventing gentrification and asserted that the new shops and market-rate rentals envisioned by the plan would further the displacement of existing residents.[31] Many also viewed the plan's clause committing development to 60 percent social housing as not nearly as significant as the city claimed. Critics asserted that the plan's broad definition of social housing meant that it would not secure housing for those in very high need, nor would it lead to the construction of as many affordable housing units as promised.[32] Furthermore, even for those low-income residents who would be able to remain, some argued, they would be unlikely to feel welcome in their own neighborhood once it had been transformed by an influx of middle-class residents and boutiques. What city planners saw as an ambitious effort to balance competing interests and secure affordable housing, others saw as an attack on the social sustainability of the Downtown Eastside.

GREEN CITY AMBITIONS AND A SUSTAINABLE SPEED OF CHANGE

From the perspective of housing justice, one might be much more concerned about the displacement of the Downtown Eastside's low-income residents than about homeowners enriched by rapidly rising housing prices who resist new multifamily properties in their neighborhood. Political power matters, and homeowners in Vancouver, as elsewhere in North America, have long proved to have outsized power in local development politics.[33] Yet there are similarities across these three cases that are worth contemplating. In each of these controversies, neighborhood groups accused the mayor and city council of forcing more and different kinds of new development on them than they wanted. Interestingly, officials did not contest this. Indeed, every city official I talked with stated that neighborhoods needed to be pushed to take on their fair share of the increased housing supply necessary to meet the demands of future population growth, enable more affordable housing, and reduce carbon emissions. For officials, neighborhood-directed planning alone cannot determine the direction of development. Achieving Vancouver's ambitious environmental and social goals requires that all neighborhoods change, perhaps more than some residents would like. As they saw it, the decisions being made regarding new developments, building standards, green space, and so on would have century-long consequences and must be made to reflect this long time horizon.

Yet some elected officials also recognized that the speed of change was potentially problematic, even if they thought the changes being made were necessary.[34] In a press conference announcing the decision to pull the Grandview-Woodland plan for further review, for example, Mayor Robertson said that they had moved "too fast" on the tower proposal and densification of this area and that this speed (but, he implied, not the towers themselves) had been a mistake.[35] Councilmembers echoed this sentiment. Speaking of Vision's agenda in general, one told me, "We are doing an enormous amount for a local government. Too much? No. But . . . we have to proceed with a little bit of caution, because . . . this pace of change is really frightening to people."[36] For her, it was not that the proposed development projects were bad ideas or that too much development was being promoted. Rather, as she saw it, it was the speed of change that garnered fear and resistance from residents.

Some community activists shared this overall assessment, asserting that some of the resistance to new development came from the fact that people simply did not want change. Residents, they told me, are deeply attached to Vancouver, particularly their own neighborhoods. As such, rapid development can threaten the sense of place and attachment people have to their communities. As one Grandview-Woodland activist said, "We're not stupid, we know neighborhoods change. But we think . . . that some places need to change less than others."[37] There is undoubtedly an important case to be made for this. Place and community matter. But as these development controversies show, who gets to decide which places "need to change less than others," and how they should change, is not a straightforward task. A senior planner articulated this challenge as follows:

> Trying to get people to understand that larger citywide, regional, global perspective, and how it plays down onto their block is a challenge. And things are getting more complex, they are not getting easier. And I think communication is a big part of it, and it's a really difficult thing to do. You know, when you are in a hot room, and you are trying to explain why social housing needs to occur in this wealthy neighborhood. Because every neighborhood has to have a variety of housing types and everyone agrees in principle, but when that social housing arrives next door it becomes an issue. Or in the Downtown Eastside the opposite, you know? Social mix is important [but low-income residents say] "We don't want [richer people] here, we don't feel comfortable around

> them, they judge us." Well, but we've got to find the right place for social mix in this neighborhood. . . . And I know people have the thought that we don't listen, but [the Downtown Eastside] plan is entirely based out of . . . what we heard from the community. Now, there are some things that we simply had to do, like "you know what? I'm sorry, we do need to have more rental housing, so we're gonna have to figure out where to put it in this area, because this talk is getting old. We do need to have some market housing [because] we have no money to pay otherwise for the social housing in this area." So, we do have to bring those sorts of imperatives to the room.[38]

BECOMING A JUST, GREEN CITY

Since the interviews included in this chapter were conducted in the mid-2010s, much has changed in British Columbia and Vancouver. In 2016, British Columbia established a tax on foreign buyers and for vacant buildings, with the hopes of discouraging the speculation that is seen as exacerbating the region's affordability crisis. In 2019, the new mayor, who was elected in 2018 when Robertson decided not to run for reelection, was working to build more new high-density buildings that include an even greater percentage of market and affordable rentals.[39] Perhaps most importantly, that same year, the BC government announced that after decades of virtually no provincial funding, it would invest $7 billion over the next decade to facilitate the construction of 114,000 units of affordable housing of various kinds across the region.[40]

These efforts may eventually have an impact. As of this writing, however, although the increase in housing prices has cooled somewhat since the mid-2010s, housing is still incredibly expensive and remains a major, unresolved challenge in Vancouver and across urban Cascadia.[41] Politicians, academics, and local commentators continue to debate the causes and best solutions to the area's affordability crisis. New efforts to incentivize developers to construct more rental properties remain controversial. Some argue that incentives excessively subsidize developers without providing enough affordable rental units in return.[42] In addition, wealthy communities continue to decry new development overall and the changes to their neighborhoods it will bring.[43]

In late 2019, the City of Vancouver announced the launch of a new citywide master planning process. It is being promoted as "much more than a land-use plan—it will address vital areas such as reconciliation, equity, climate change, affordability, housing, the economy, jobs, transportation, cultural vibrancy and neighbourhood design."[44] This is a tall order, and undoubtedly the deep disagreements about the location, pace, and type of growth that should occur in Vancouver, let alone who benefits from this development, will have a predominant role to play in this process.

Meanwhile, Vancouver continues to be a leading green city, on track to achieve many of the environmental and climate goals it adopted a decade ago.[45] Despite substantial growth in population and economic activity, overall carbon emissions have decreased. Emissions from municipal operations have more than halved. Green transportation use hit 50 percent in 2015, facilitated by the construction of robust cycling and transit networks. Local food assets and green jobs have grown, solid waste sent to landfills and incinerators has declined, and new buildings have half of the carbon footprint they did in 2007.[46] In early 2019, Vancouver declared a climate emergency, accompanied by a detailed plan to reduce carbon emissions across the city even further.[47] As North American cities strive to address climate change and their overall environmental impact, Vancouver remains one of the most ambitious and successful at doing so.

However, just climate resilience and participatory justice require that communities have control over the conditions of their lives.[48] The development controversies in Vancouver show how hard this is to achieve in practice. How does a city reduce its environmental impact while increasing access to affordable housing for all? What criticisms should rightly be raised against the imperative city leaders see to approve ever more high-rise towers in order to "claw back" from developers the accompanying increase in land value for community amenities and affordable housing? Can the city go further than it has to push developers for more affordable rental homes and to protect low-income residents? In a city that is already quite gentrified, how can one tell the difference between NIMBYism and legitimate concern about new development not adequately respecting the preferences and connection to place of existing residents? How can community resistance to change be reconciled with the swift, significant transformations in built form, living patterns, and transportation modes that must occur for greenhouse gas emissions to be cut 80 percent by 2050?

The form of cities needs to be rethought to create urban areas that are healthier, are more just, and have a smaller environmental impact in order to reduce and prepare for the risks presented by climate change. Yet the development controversies in Vancouver offer an important lesson to those cities, including the other major cities of Cascadia, that are engaging in ambitious efforts to become more environmentally sustainable. Although the need to mitigate and adapt to a changing and threatened environment is great, and although cities do hold an important role in furthering these efforts, urban environmental initiatives must be undertaken in a way that also strengthens the social fabric of the city.

Of course, there is no clear formula that defines a socially sustainable rate of change or that outlines how to balance the urgency of creating more resilient cities for the long term while, in the short term, not disrupting people's connections to community and place. Crucial to finding this balance will be a holistic approach that wrestles explicitly with the intersection of a socially and environmentally sustainable city, of continuity and change. Vancouver's willingness to set challenging environmental goals and then take concrete steps to reach them is laudable. But Vancouver leaders must do more to establish a shared vision around the changes necessary to achieve these goals. Meanwhile, Vancouver residents must continue to hold officials accountable to local concerns and to the importance of people's shared sense of community and place. As Vancouver continues toward greater density and greener buildings, as residents demand a continued voice in the planning process, and as the provincial government begins to engage more actively in affordable housing provision, it is possible—though by no means certain—that Vancouver can become a model for Cascadian cities' movement toward justice and sustainability.

NOTES

1 Casey Jaywork, "Three Cities, One Housing Crisis," *Cascadia Magazine*, January 29, 2018.

2 Emily Badger and Keven Quealy, "Watch 4 Decades of Inequality Drive American Cities Apart," *New York Times*, December 2, 2019; and Richard Florida, *The New Urban Crisis* (New York: Basic Books, 2017).

3 Iain Marlow and Brent Jang, "Vancouver's Real Estate Boom: The Rising Price of 'Heaven,'" *Globe and Mail*, October 10, 2014; and Kazi Stastna, "Real Estate Rules Don't Discriminate against Foreigners," *CBC News*, March 19, 2012.

4 Noah Quastel, Markus Moos, and Nicholas Lynch, "Sustainability-as-Density and the Return of the Social: The Case of Vancouver, British Columbia," *Urban Geography* 33, no. 7 (2012): 1055–84; Corina McKendry, *Greening Post-Industrial Cities: Growth, Equity, and Environmental Governance* (New York: Routledge, 2018); and Jennifer L. Rice et al., "Contradictions of the Climate-Friendly City: New Perspectives on Eco-gentrification and Housing Justice," *International Journal of Urban and Regional Research* 44, no. 1 (2020): 145–65.

5 Aiden While, Andrew E. G. Jonas, and David Gibbs, "The Environment and the Entrepreneurial City: Searching for the Urban 'Sustainability Fix' in Manchester and Leeds," *International Journal of Urban and Regional Research* 28, no. 3 (2004): 549–69; and Corina McKendry and Nik Janos, "Greening the Industrial City: Equity, Environment, and Economic Growth in Seattle and Chicago," *International Environmental Agreements: Politics, Law and Economics* 15, no. 1 (2015): 45–60.

6 "Where the Livin' Is Easiest," *The Economist*, February 21, 2011; Tim Kiladze, "World's Most Beautiful Cities," *Forbes*, January 22, 2010; and City of Vancouver, *Greenest City 2020 Action Plan* (Vancouver: City of Vancouver, 2012).

7 John Punter, *The Vancouver Achievement: Urban Planning and Design* (Vancouver: UBC Press, 2003).

8 "Vancouver's Housing 2nd Least Affordable in World," *CBC News British Columbia*, January 21, 2014.

9 McKendry, *Greening Post-Industrial Cities*; Nik Janos and Corina McKendry, "Globalization, Governance, and Re-naturing the Industrial City: Chicago, IL and Seattle, WA," in *The Power of Cities in International Relations*, ed. Simon Curtis (New York: Routledge, 2014).

10 McKendry, *Greening Post-Industrial Cities.*

11 City of Vancouver, *Greenest City 2020 Action Plan: 2017–2018 Implementation Update* (Vancouver: City of Vancouver, 2018).

12 Lance Berelowitz, *Dream City: Vancouver and the Global Imagination* (Vancouver: Douglas & McIntyre, 2005).

13 City of Vancouver, *EcoDensity: How Density, Design, and Land Use Will Contribute to Environmental Sustainability, Affordability, and Livability* (Vancouver: City of Vancouver, 2008), 1.

14 Marit Rosol, "Vancouver's 'Ecodensity' Planning Initiative: A Struggle over Hegemony?," *Urban Studies* 50, no. 11 (2013): 2238–55.

15 Rosol, "Vancouver's 'Ecodensity' Planning Initiative."

16 Personal communication with the author, August 2014.

17 Grandview-Woodland community activist, personal communication with the author, August 2014.

18 Grandview-Woodland community activist; and Charles Campbell, "At Ground Zero for Vancouver's Towering Debate," *The Tyee*, June 25, 2013.
19 Personal communication with the author, August 2014.
20 Jak King, "Grandview Woodland Community Plan (2016) Drives Land Assembly, Price Escalation," CityHallWatch, September 8, 2017, https://cityhallwatch.wordpress.com/2017/09/08/grandview-woodland-plan-prices-up/.
21 Justin McElroy, "One Chart Shows How Unprecedented Vancouver's Real Estate Situation Is," *Global News*, February 21, 2016, https://globalnews.ca/news/2531266/one-chart-shows-how-unprecedented-vancouvers-real-estate-situation-is/.
22 City of Vancouver, *Marpole Community Plan* (Vancouver: City of Vancouver, 2015).
23 Jeff Lee, "Neighbourhoods in Revolt," *Vancouver Sun*, September 24, 2013.
24 Grandview-Woodland community activist, personal communication with the author, August 2014.
25 Personal communication with the author, August 2014.
26 David Ley and Cory Dobson, "Are There Limits to Gentrification? The Contexts of Impeded Gentrification in Vancouver," *Urban Studies* 45, no. 12 (2008): 2471–98.
27 City planner, personal communication with the author, August 2014.
28 City of Vancouver, *Downtown Eastside Plan* (Vancouver: City of Vancouver, 2014).
29 Personal communication with the author, August 2014.
30 Nathan Edelson and Karen O'Shannacery, "Revitalizing the Heart of the City," *The Tyee*, March 13, 2014.
31 Ivan Drury, "Vancouver Rightists Rising against the Downtown Eastside," The Mainlander, March 10, 2014, https://themainlander.com/2014/03/10/vancouver-rightists-rising-against-the-downtown-eastside/.
32 Maria Wallstam and Nathan Crompton, "Social Housing without Guarantees: The Myth of the 60/40 Ratio in the DTES Oppenheimer District," The Mainlander, March 12, 2014, https://themainlander.com/2014/03/12/social-housing-without-guarantees-the-myth-of-the-6040-ratio-in-the-dtes-oppenheimer-district/.
33 David Ley, *The New Middle Class and the Remaking of the Central City* (Oxford: Oxford University Press, 1996).
34 City councilor, personal communication with the author, August 2014.
35 Jeff Lee, "Party Moved 'Too Fast' on East Van Towers Bid," *Vancouver Sun*, May 5, 2014.
36 City councilmember, personal communication with the author, August 2014.
37 Grandview-Woodland activist, personal communication with the author, August 2014.

38 Personal communication with the author, August 2014.

39 Frances Bula, "Affordable Housing in Vancouver Can Only Happen under New Rental Program, Stewart Says," *Globe and Mail*, April 14, 2019.

40 British Columbia, Municipal Affairs and Housing, "BC Government Addressing Housing Affordability Challenges," accessed July 22, 2019, https://news.gov.bc.ca/factsheets/bc-government-addressing-housing-affordability-challenges.

41 Shawna Kwan, "Hong Kong Housing Is World's Least Affordable for 9th Year," Bloomberg News, January 21, 2019.

42 Matt Robinson, "Councillors, Developers at Odds over Affordable Rental Policies in Vancouver," *Vancouver Sun*, March 25, 2019.

43 Bula, "Affordable Housing."

44 City of Vancouver, *Greenest City News* (newsletter), December 2019.

45 City of Vancouver, *Climate Emergency Response* (Vancouver: City Planning Department, 2019).

46 Vancouver, *Greenest City News*.

47 Vancouver, *Climate Emergency Response*.

48 Siri Eriksen et al., "When Not Every Response to Climate Change Is a Good One: Identifying Principles for Sustainable Adaptation," *Climate and Development* 3, no. 1 (2011): 7–20; and David Schlosberg, "Climate Justice and Capabilities: A Framework for Adaptation Policy," *Ethics & International Affairs* 26, no. 4 (2012): 445–61.

CHAPTER 11

ECOLOGICAL DEMOCRACY AND THE DUWAMISH RIVER CLEANUP

MARK PURCELL

LET ME START WITH A PROVOCATION: STOP TALKING ABOUT justice so much and start talking about democracy. This chapter elaborates that provocation through an investigation of the Duwamish River Cleanup Coalition (DRCC), one of the many movements in Cascadia that are struggling to resist and rethink the capitalist urbanization of nature. When those who seek alternatives to the capitalist urbanization of nature think too much in terms of justice, I argue, we are tempted to become obsessed with outcomes, rather than the process of politics, and we tend to fall into the trap of becoming passive recipients of outcomes that are delivered to us by others. However, when we begin to think in terms of democracy, we become alive to the possibility of our own political activation. We begin to see that it is vitally important that people take their affairs into their own hands, that they produce for themselves the outcomes they desire. More specifically, when we think in terms of a particularly ecological democracy, we are able to radically expand our idea of democracy. It is not only humans that make up the *demos* of democracy, but a whole range of nonhuman entities as well.

The chapter examines the DRCC not in order to present an ideal example of ecological democracy. Rather, the DRCC is important, I think, because they grapple with the same issues of justice and democracy this chapter explores. The DRCC, I argue, thinks often in terms of justice, but they are also able to conceive of their work in terms of democracy. To the extent they do the latter, their work proposes to us the possibility of an ecological democracy. It is a possibility we would do well to be attentive to.

WHAT IS DEMOCRACY?

To defend my claim that we need to begin thinking more in terms of democracy, I need to say a little about how I understand democracy. I do *not* mean democracy as it is conventionally understood: a form of government in which people vote periodically so that they have some say in selecting their leaders. Instead, I understand the term more literally: democracy is when *demos* (people) retain their *kratos* (power) and use it to manage their affairs for themselves.[1] Democracy is a way of living in which *demos* and *kratos* are never separated; people never surrender their power to an entity that is other than themselves. Their power remains always in their possession, and they use it together with others to manage their collective affairs. In a typical liberal-democratic system like the United States, the act of voting is precisely this surrender of power: people agree to transfer their power to another person—a senator or mayor or president, and so on—and that other person wields their power on their behalf. We call this system democracy, but it is precisely the opposite. Democracy is when people manage their collective affairs themselves.

The State is the primary example of an entity that is other than people, but so too is the capitalist corporation. In democracy people manage not only their political affairs but also their *economic* affairs themselves. Democracy is a way of living together without, or beyond, both the State and capitalism.[2] This idea of democracy might appear quite radical and therefore impractical. How could people manage their affairs *entirely* themselves? That is not an unreasonable question, and it prompts me to be clear that democracy should not be understood to mean a perfect community, a utopia at the end of history that we expect to achieve in the near future. Instead, democracy is best understood as a *project*, as a patient, insistent, and long-term project to become active and increasingly manage our affairs ourselves. Now, of course, such a project will necessarily confront this question of community: Who are "we," the *demos*, what relations should we have to each other, what kind of community do we want to create together? Democracy is not inclined to answer such questions in advance. It requires people to work on those questions themselves. The democratic project insists only that people manage their affairs for themselves and that they do so as equals: each person is assumed to be equally a part of the community and equally qualified to participate in its collective affairs.[3]

WHAT'S WRONG WITH JUSTICE?

To be clear, I don't mean this chapter to say that we should never think in terms of justice. Such thinking is, broadly, a good thing. For example, the movement for "environmental justice" has increased awareness of, and resistance to, the consistent pattern of poor neighborhoods and communities of color being exposed to far more environmental hazards and facing far worse health outcomes than rich or white neighborhoods.[4] What I am saying is that those on the Left who desire a radically different city, whether it be ecologically, economically, or politically, almost always think in terms of justice, and we rarely think in terms of democracy. This habit of thought limits us, and it limits us, in particular, from the point of view of democracy. When we think in terms of justice, we are led very strongly in the direction of focusing on particular outcomes. Environmental justice, for example, has a long tradition of struggling for a more equitable distribution of the environmental risk that capitalist urbanization produces. This *outcome* very often becomes the essence of the political agenda of the movement. All thought and action is consumed by it. Any effective means, whatever their potential drawbacks, will be employed to achieve it. If progress toward this outcome is not achieved, the movement will be considered a failure.[5] When we think mostly in terms of justice, we are in danger of missing another critical component of any political movement: the activation, the engagement, the learning, and the growth among people who participate in it. From the perspective of democracy, in which the whole point is for people to become active and take their affairs into their own hands, such activation and growth is absolutely central. The point of a movement for democracy is for people to increasingly realize their own power to do things for themselves, to develop that power, and to come to prefer keeping that power instead of surrendering it. For democracy, the specific outcomes in this round of political struggle—the siting of a waste facility, for example, or even the more general achievement of a law requiring more equitable distribution of hazards—are important, to be sure, but they are secondary to the question of the development of people's desire, capability, and power.

Another danger when we think in terms of justice, and we value the just outcome above all other considerations, is that we are very likely to be willing to trade democracy for justice. This occurs when, for example, a strong leader in the movement is able to "make things happen," to deliver the outcomes the movement desires. This leader might be effective because she is charismatic, or closely connected to State power, or able to marshal financial resources. And if the purpose of the movement is to deliver just outcomes, and this leader's activity is the most effective means to achieve those outcomes, then people will, of course, choose to surrender their power to the leader, to let her take care of it. This path will deliver justice, but it will directly weaken the democratic life of the movement. It will teach people that a good movement requires strong leaders who will act—instead of people—to deliver good results. It will deactivate the population and convince them they are better off remaining passive and surrendering their power to an effective leader. From the point of view of justice this is fine, but from the point of view of democracy it is a disaster.

JUSTICE AND DEMOCRACY IN THE DUWAMISH RIVER CLEANUP

In this section I ground these somewhat abstract questions of justice and democracy in the actual practice of the Duwamish River Cleanup Coalition. I don't present the DRCC as an example of a movement that thinks in terms of democracy rather than justice. Indeed, overall the DRCC is probably more inclined to think in terms of justice. But their activism is, I think, a rich ground for exploring these questions, because they exhibit both a strong desire for justice and a (perhaps less) strong desire for democracy. And they imagine both ideas to be always closely linked to the question of ecology. Let me begin by giving some background on the group and then examine their political desires in more detail.

A BRIEF HISTORY

The Duwamish River is Seattle's main river. It is formed by the confluence of the much longer Green and the Black Rivers, and it empties into Elliott Bay just south of downtown Seattle. Before 1913, the

Duwamish meandered through a flat floodplain and created a rich estuary in the river's delta, which emptied into Elliott Bay. In 1913, the US Army Corps of Engineers dredged and straightened the river to make it more functional for the industrial economy and the new Port of Seattle. This process replaced the Duwamish River with something that is now called the Lower Duwamish Waterway. Since its inception, the Waterway has been lined with industrial production facilities such as Boeing's Plant 2, which made the B-17 bomber during World War II (see chapter 2).

This long history of heavy industrial manufacturing, not surprisingly, steadily polluted the Lower Duwamish Waterway, and in 2001, the US Environmental Protection Agency (EPA) declared the Lower Duwamish a Superfund site. Superfund is an EPA program that determines the most polluted sites in the country and sets out a detailed bureaucratic process to carry out their cleanup. The idea of Superfund is that those who are responsible for the pollution, the potentially responsible parties (or PRPs), should both pay for and carry out the cleanup. In this case the four primary PRPs were Boeing, the Port of Seattle, the City of Seattle, and King County. The PRPs joined together to form the Lower Duwamish Waterway Group (LDWG), and that organization has been primarily responsible for carrying out the cleanup. It has measured and mapped the pollution, devised cleanup plans, and funded and carried out those plans.[6]

Part of the law that governs Superfund requires that "the community" have an effective voice in the decisions concerning the cleanup. It requires the creation of what is called a Community Advisory Group to be the official representative of the community. The DRCC formed originally to serve as the Community Advisory Group for the Duwamish Superfund cleanup. Its members comprise a variety of community groups including the Duwamish Tribe, the Indigenous people who traditionally inhabited the Duwamish River; several influential environmental advocacy organizations; neighborhood associations in South Park and Georgetown, the two Seattle neighborhoods most proximate to the cleanup; an environmental justice advocacy organization; and an association of small businesses in the immediate area.

The Superfund cleanup is a very long process and proceeds generally in the stages I mentioned above: study, plan, cleanup. The Duwamish cleanup had an initial phase (the first ten years or so) in which several "hot spots" were identified, areas in the river where pollution was particularly acute, and these were studied and cleaned up first by various members of LDWG. When the hot spots were completed, the more general cleanup began, and the EPA finalized the plan for that in 2014. Throughout this long process, the consistent dynamic, for both the hot spots and the making of the general plan, has been that LDWG proposes relatively sparse and low-cost plans to the EPA, and the DRCC lobbies the EPA on behalf of the community for more comprehensive and more expensive action.

In addition to this advocacy work inside the Superfund process, the DRCC has also developed, over time, into an organization that takes action beyond the confines of Superfund to improve the condition of life in the wider Duwamish Valley. It has formed partnerships with a variety of organizations to pursue small grants that enable it to undertake projects such as restoring habitat along the river, providing environmental amenities in South Park and Georgetown, conducting environmental health studies, training youth for "green jobs," and hosting festivals to celebrate the river's past and present. So while the DRCC is first and foremost a Community Advisory Group to the EPA for the Superfund cleanup, it is also many other things besides: a community organizer, a research institute, an environmental education organization, an ecological restoration nonprofit, and more.

THE DUWAMISH RIVER CLEANUP COALITION DESIRES JUSTICE . . .

As I mentioned above, the DRCC is particularly salient to this chapter because it desires both justice and democracy. The former desire is most apparent in its work inside the Superfund process. Until 2020, the DRCC's mission statement made clear that they sought, primarily, to ensure a certain cleanup outcome: "Our mission is to ensure a Duwamish River cleanup that is accepted by and benefits the community and protects fish, wildlife and human health."[7] At each hot spot, for example, LDWG proposed a cleanup that met minimal legal standards, which the EPA could therefore legally approve, but then the DRCC pressed EPA to require more than the minimum. In the case of one hot spot called Terminal 117, for example, LDWG proposed a standard cleanup of the river sediment around the site, and the DRCC asked for, and ultimately secured, further testing of an upland area near houses in Georgetown where the soil was contaminated with polychlorinated biphenyls (PCBs). The DRCC used several different tactics in this effort. They secured a grant to hire their own scientific expert, who critiqued as inadequate the pollution sampling procedure LDWG used; they made alliances with pro-environment elected officials, who lobbied on their behalf; and they packed the public meetings on the plan with local residents demanding that LDWG do a more thorough job cleaning up their mess. In this case, the EPA decided to require further study, PCBs were found in the upland area, and the cleanup of that area was added to the hot spot plan.[8]

This outcome is precisely the kind of politics the desire for justice produces: the DRCC pressures the EPA to require LDWG to deliver a more robust cleanup outcome to the community than they otherwise would have. The point of such politics is to achieve outcomes, by whatever means, that increase justice, however defined. In this case, the DRCC used a variety of means inside the Superfund process to secure more justice, understood to mean a more extensive and expensive cleanup that directly benefited local residents. For most in the DRCC, their understanding of justice was rooted in an idea of environmental justice, understood in the conventional way: the historical distribution of environmental hazards—such as PCBs—and their resulting health problems are far more prevalent in poor neighborhoods with a high percentage of people of color, such as South Park and Georgetown, than they are in rich, white neighborhoods. Struggling for a more thorough cleanup, then, is a way to redress this unjust history. This idea was explicitly articulated by community members at the public meetings for Terminal 117. Specifically because of a history of environmental injustice, they argued, LDWG needed to do *more* than the minimum allowable cleanup to begin to make up for the injustices of the past.[9] In this same vein, during the process to produce the general cleanup plan for the Duwamish, the DRCC successfully lobbied the EPA to do the first environmental justice analysis in Superfund history.[10] And beyond the cleanup, the DRCC partnered with the School of Public Health at the University of Washington and a nonprofit called Just Health Action to produce health impact studies that showed the zip codes in which South Park and Georgetown sit to be disproportionately unhealthy and have much lower life expectancies.[11]

So we can see a very strong desire for justice, understood particularly as environmental justice, animating the DRCC's work. This desire impels the leadership of the DRCC to work on behalf of the community to secure cleanup outcomes that will most benefit the community. According to this way of thinking, these outcomes are the substance and the measure of the DRCC's work. The more robust the cleanup outcomes, the more successful the DRCC is. For example, during the process of deciding on the general cleanup plan, the DRCC launched a social media campaign, called #riverforall.[12] The idea was to build a groundswell on social media that would raise awareness of the EPA's impending decision and encourage people to contact the EPA and pressure them to adopt a more robust plan. In the end, this campaign had an effect. As the DRCC put it, #riverforall "encouraged everyone to see the Duwamish River as our river. Because of this effort, more than 43,000 people from all 50 states in the nation asked the City of Seattle and King County for a more comprehensive, health-protective river cleanup."[13] This dynamic lies at the core of the justice approach: people are activated, but only in order to ask the government to deliver a better outcome to them. This relation is reiterated in the DRCC's fact sheet on the final cleanup plan: "In the coming months and years, we will be helping you talk to your local elected officials about taking action to support additional cleanup to ensure that our river is clean and healthy for future generations."[14] Here we see the justice approach in its least democratic form: the community plays the role of supplicant to the State, asking it to deliver what they want. Moreover, in this instance the DRCC imagines people to be incapable of even this supplicant role, saying they will "help" people talk to their elected officials. In this vision, people are basically inactive by default, and they need to be activated by an organized community leadership. What is more, this activation is only useful in order to plead with those in power to deliver better outcomes.

. . . AND DEMOCRACY

A democratic imagination does not conceive of politics this way. It sees the activation of people not as a means to achieve the end of a more robust cleanup, but as an end itself. It does not start from the assumption that people are inactive and they need a vanguard group to activate them. It instead assumes that people are already active, and capable, and it seeks only to enhance those activities and capabilities. Democracy refuses the assumption that people must appeal to State power to improve the river; it searches tirelessly for ways that people can take up that task themselves. Democracy does not insist that people always deliver better outcomes than the State, although they very often do. It insists, rather, that those outcomes are not the primary goal of politics. Politics, instead, is a project in which people become active and alive, in which they insistently increase their ability to manage their affairs for themselves.

This democratic way of seeing politics is not alien to the DRCC. While they strongly desire justice, they also desire democracy. That is, they work to help people become ever more active in order to take the future of the river, and its valley, into their own hands. In this case "people" refers especially, but not only, to the DRCC's particular constituencies: Indigenous people, neighborhood inhabitants (especially people of color and immigrants), environmental activists, and small businesses in the area. I think it is fair to say that this desire for democracy is more fledgling and rare in the DRCC's work, but it is very much there, and it is something we should be attentive to.

Perhaps the most obvious example of the DRCC's desire for democracy is their tireless work to transform the EPA's anemic vision of public participation into something far more robust. US Superfund law is heir to the Great Society reforms of the 1960s that required federal decision-making to meaningfully involve community members. DRCC's very existence, as the Community Advisory Group to the Superfund cleanup, is a result of such requirements. But the actual practice of the EPA is what most observers of urban politics would consider a fairly typical setup: powerful stakeholders work with the government to develop a plan of action (and sometimes one alternative plan), and then they propose that plan to "the community" and ask for their feedback. Sometimes this feedback is given at public meetings, but increasingly it is offered online. Sometimes the feedback is intelligent and helpful, often it is not. Sometimes the feedback prompts minor adjustments to the original plan, usually it does not. The DRCC has sought to improve this process in multiple ways. They have secured grants, available from the EPA, to hire an independent scientific expert who provides testimony that can critically question LDWG's scientific studies. They have worked to expand the opportunity for the community to offer input into cleanup decisions. They have lobbied to be at the table while the draft plans were being made, rather than only giving input once the draft plan was completed. They have also worked insistently to make the comments offered at public meetings more effective. For example, before the public hearings on each hot spot, they met extensively with community members to develop a vision for the Duwamish Valley in general and, using that vision, produce a set of coherent suggestions for a given cleanup plan.[15] Then, at the public hearing for that plan, instead of the string of uncoordinated comments that are typical of such public meetings, people articulated a consistent argument about why the draft plan fell short and how it could be improved.[16]

I do not want to overstate the DRCC's success here. The EPA's institutional machinery has pretty effectively tamed the requirement for meaningful public input in Superfund. Moreover, one could argue that what the DRCC is really doing here is merely coming up with a more effective strategy to achieve better justice outcomes. But I think it is more true to say that in this effort to enhance EPA's process, the DRCC desires *both* justice and democracy. Enhanced public participation is a way to increase justice outcomes, but it is also a way to activate and mobilize people. It is an opportunity to bring them together to learn about the river and the cleanup, to meet each other, and learn about each other, to discuss their vision together, to decide collectively on a strategy to realize that vision, and to put that strategy into practice.

This way to read the DRCC's desire, as a desire for justice but also for democracy, is corroborated by the many other initiatives the DRCC has undertaken, which I mention above, initiatives that range beyond the confines of the Superfund cleanup. The #riverforall campaign, while it was partly a way to put public pressure on EPA for a more just outcome, was also a way to get the word out about the Duwamish River, to encourage people in the valley to connect to the river, to take responsibility for it, and to begin to see the river as their own. The website for the campaign, www.riverforall.org, featured a series of community members, including the famous hip-hop artist Macklemore, saying "The Duwamish is my river." Along these same lines, the DRCC has long been providing tours of the river—on foot or more famously by kayak—that help participants learn about the river's history and ecosystem. The DRCC also hosts an annual Duwamish River Festival, which similarly provides city residents an opportunity not only to learn about the river, but to celebrate it, and invites them to affirm a commitment to take care of it.

This educational function is also provided by the DRCC's participation in a coalition called Duwamish Alive! that organizes habitat restoration projects. These small-scale work parties are an opportunity to do more than just learn about the river; instead, people perform ecosystem protection, management, and restoration activities together. While these are small actions, probably less ecologically important as a whole than the Superfund cleanup process, they are nevertheless numerous and significant interventions along the river that operate independently of EPA and Superfund.[17] While there are governmental partners in the coalition, Duwamish Alive! is mostly a community conceived and driven initiative. Another DRCC project in this vein is their Duwamish Valley Youth Corps, a job-training program for local teens that prepares them for careers in environmental restoration and community organizing. The DRCC is also a key collaborator in a number of other projects to improve the local environment in low-income neighborhoods in the Duwamish Valley, including tree canopy restoration, a "Green Wall" to minimize localized pollution effects, efforts to increase open space and park acreage, and initiatives to reduce air pollution in the area.

In addition to all these activities, the DRCC has also led a long-term public visioning process, as we saw above. This effort brings people in the Duwamish Valley together, in many events over years (often in the style of a design charrette), to discuss what people desire for the valley and articulate that vision as a formal plan. Of course, this plan has been mobilized inside the Superfund process to leverage a more robust cleanup, but its energy has also been directed in other directions beyond the cleanup, to projects conceived and led by the DRCC and community members, many of which I have mentioned in the paragraphs immediately above. The main concern that arose in this visioning process had to do with health broadly conceived: human health and sickness as a result of air, water, and soil pollution, but also the ecological health of the valley. The health study we saw above, which found clear patterns of citywide inequality, was one result of this process. Another project was a community-based effort to identify and address air quality problems in the valley. In this case the DRCC won a grant through the EPA's Environmental Justice Collaborative Problem-Solving Cooperative Agreement. They collaborated again with Just Health Action, as well as the American Lung Association, to identify very acute localized air pollution dangers and provide specific solutions such as air filters to affected households.[18]

To be sure, none of the initiatives in this section are democracy in its purest form. I think they are better seen as evidence of an emerging democratic desire within the DRCC in the context of a predominant desire for justice. People in the Duwamish Valley are becoming active in order to deliver more just outcomes, but they are also, at least in part, becoming active for its own sake. Even though, in many cases, the DRCC is often operating under the auspices of government programs, and appeals to the State for all sorts of assistance, there is a sense in which they are aiming at building an active community that does not need to rely so heavily on the State. The DRCC still serves very much as a vanguard, as a core leadership group that acts on behalf of the community, but they also, sometimes, are working to create the conditions that will make their leadership no longer necessary, because people have become engaged to the point that they are able to act on their own behalf.

ECOLOGICAL DEMOCRACY?

Being attentive to the DRCC's work helps open up this more democratic way of thinking, but it also helps to push this same thinking beyond its conventional ambit. That is, it allows us to begin to think in terms of what we might call an ecological democracy. In the description of democracy I give above, and throughout the chapter, the *demos* of democracy is very much understood to mean people, which is to say human beings. Moreover, these people, in the root idea of democracy, are not just any people but city dwellers: inhabitants of a *polis*. These associations—with the human and the urban—are very deeply rooted in the history of democracy. But it is not so far-fetched, I think, to push beyond these limits, to conceive of the *demos* more broadly to include nonhuman species and nonurban places. We might extend the meaning of *demos*, in other words, to mean all the biophysical elements, both alive and not, of the entire ecological system, both urban and not, in which a community lives and on which it depends. This ecosystem could be very small, something as small as the Duwamish Valley, but, of course, ecosystemic thinking would insist that any small system is connected to and dependent on larger systems, at every scale all the way up to the Earth.

To see how this ecological extension of democracy would look in the context of the Duwamish, it is instructive to examine the names of the two primary groups I have been discussing: the DRCC and LDWG. The DRCC is the Duwamish *River* Cleanup Coalition, while LDWG is the Lower Duwamish *Waterway* Group. It might seem like semantics to dwell on this difference, but it is in fact quite meaningful. As we saw, the Duwamish River was officially renamed the Duwamish Waterway in 1912 when the Army Corps straightened and channelized the river to facilitate the construction of the port and jump-start the industrial economy. Whatever its benefits, this was also an act of great violence. It sacrificed the riverine ecosystem to the needs of the economy, and it drove home the colonial dispossession of the Duwamish Tribe, whose life and livelihood were tightly bound up with that ecosystem (see chapter 5).[19] Insisting on the term *waterway*, as LDWG does, and as EPA does, is a reinscription of that violence. Insisting, instead, that the Duwamish is a river is both a critique of 1912 and a proposal for a radically different vision for the valley.

Throughout the Superfund process and in their other activities, the DRCC insistently repeats the claim that the Duwamish is a river, not a waterway, and that the Superfund process should be reimagined as one that might fully restore the Duwamish riverine ecosystem, rather than merely clean up the pollution in the Duwamish Waterway to a level required by law.[20] This is a quite profound demand, one that envisions reversing (not mitigating) both the century-long sacrifice of ecology to economy and the colonization and dispossession of the Duwamish Tribe. The DRCC are under no illusions that the Superfund process will actually deliver such an outcome, and indeed the EPA and LDWG have shown themselves to be spectacularly unable to imagine anything of the sort. But in imagining such an alternative, and in articulating it consistently, the DRCC has steadily opened up the possibility of both radically extending the project of Superfund and radically rethinking democracy.

What would that mean, to conceive of an ecological democracy in the Duwamish Valley? Answering that question fully is far beyond the scope of this chapter, so let me just offer a quick sketch of one possible vision. Ecological democracy in the Duwamish Valley would imagine a *demos* that of course includes all people in the valley's ecosystem (irrespective of existing political jurisdictions), from the headwaters in the Cascade Mountains to the river mouth in Elliott Bay. But that *demos* would also include all nonhuman species in the system, both the sentient, such as eagles or salmon, and the insentient, such as eelgrass and Douglas fir. This *demos* would include the inanimate elements of that system as well: soil, sediment, water, air, and so on. All of it, from an ecosystemic point of view, is a vital part of the community. Democracy, in this way of thinking, would be the project whereby this new *demos* becomes active and increasingly manages its affairs for itself. This *demos* would express and consider the interests and desires of all its members, animate and not. Each member would be considered equally part of an interdependent system of agents seeking to manage their collective affairs themselves. Each would hold a value in and of themselves. The salmon would not be considered because they are a food source for humans, but because they are an equal member of the community. Eelgrass, similarly, would not be valued because it provides vital salmon habitat, but because it is, in and of itself, fully part of the *demos*. Sediment, air, water as well—all full members of the *demos*, all inherently valuable.

Of course, a *demos* that includes the nonhuman, and even the inanimate, invites practical questions. How, for example, will this *demos* communicate? In the classical understanding of democracy, humans come together in the *agora* to discuss their affairs through rational discourse (in Greek). Clearly that understanding is inadequate to ecological democracy, since the salmon don't speak Greek. Instead of rushing to conclude, therefore, that ecological democracy is impossible, we might instead begin to imagine entirely new forms of communication within the *demos*, ways that humans and salmon and eelgrass and sediment might express themselves to each other and be understood. Ecological democracy urges us to let go of the idea that communication must take the form of rational argumentation using human language. Letting go in this way would allow us to better see that, in fact, all the members of this ecological *demos* are already communicating with each other, or, better, they are all already expressing themselves. They are all constantly discharging their particular *kratos*—their power—into the system. Humans straighten and channelize the river; eagles and orcas hunt the salmon and eat them; salmon migrate, despite it all, to their spawning grounds to reproduce; eelgrass takes in nutrients and oxygenates the water; and the water flows, always down, carrying with it any number of other members, such as gases and pollutants and sediment and more besides. In a democracy like this, all members of the *demos* would need to acquire new languages, new ways of listening to and understanding each other, new ways to be attentive to each other. The eelgrass is always speaking, always expressing itself into the system, as are the salmon, and the water, and all the rest. The question is whether those expressions are being heard, and attended to, by other members of the *demos*.

Maybe this new learning would fall disproportionately on humans, who have become so very inattentive to the others in the system. In the case of the Duwamish cleanup, I think there is much to be learned from the wisdom and knowledge of the Duwamish Tribe, whose relationship with the Valley was, and is, less tone-deaf than is the dominant colonial culture. Tribal leaders have been expressing this wisdom throughout the Superfund process, but it has not had nearly the influence it should have. At the same time, we must not fall into the easy idea that we just need to return to Indigenous thought and ways of life, which are so often romanticized. The project of ecological democracy, instead, is a project into the future, a project whereby all of us in the *demos* invent, together, a *new* form of life, a new equal life in common, based on a new understanding of the *demos*, and very new forms of communication.

CONCLUSION

Let me conclude my exploration for a new ecological democracy by connecting it explicitly to the question of the bioregion of "Cascadia" that animates this book. Here I think my exhortation to think less in terms of justice and more in terms of democracy bears some important fruit. Imagining Cascadia from the perspective of justice tempts us very quickly, I think, into a bioregionalism that pursues strong government, top-down policy that redistributes risk and privilege, both social and ecological, across the whole region. Whatever benefits this policy might provide in terms of more just outcomes, it runs the risk of deactivating the population. The perspective of democracy would instead seek ways the entire bioregional *demos* (both human and nonhuman) might become active and take up the task of managing the bioregion themselves. The justice approach to Cascadia is a less radical project because it would ask us only to extend our current concept of State sovereignty to cross borders it has not yet crossed. A more radical, democratic project would seek to unlearn that concept of sovereign authority and begin the project of imagining and practicing a new, democratic form of life in the bioregion. Moreover, if we think of Cascadia in terms of a specifically *ecological* democracy, then we would have to develop this new form of life in the bioregion in concert with a radically expanded idea of what the *demos* means. This project will set about creating a new form of life, called Cascadia, in which all elements of the ecosystem are full members of the *demos*, equal in their rights and inherent value, and they invent new ways to communicate effectively with each other about their needs and desires.

I imagine that last sentence might sound crazy. You might object that it is impractical, that it will never happen. I guess what I would ask is, before you judge its practicality, ask yourself if it sounds good, if it seems like something that we need, something that would be good for the world. If it is, then let's start practicing it, together with others, today. Let's start forging a new form of ecological-democratic life together, in the Duwamish Valley, in Cascadia, and on the Earth.

NOTES

1 Mark Purcell, *The Down-Deep Delight of Democracy* (Chichester, UK: Wiley-Blackwell, 2013).

2 And so it is not at all the same thing as libertarianism as it is understood in the United States, which seeks to free people from the control of the State, only to deliver them over to capitalist corporations.

3 Jacques Rancière, *Dis-agreement: Politics and Philosophy*, trans. Julie Rose (Minneapolis: University of Minnesota Press, 1999).

4 See David Schlosberg, *Defining Environmental Justice: Theories, Movements, and Nature* (Oxford: Oxford University Press, 2007).

5 I am not a scholar of the academic environmental justice literature more broadly, and so what I am describing here is not intended to capture nuances in that literature, but rather the way that justice is conceived of commonly in EJ practice. The Duwamish case, below, presents an excellent example: their *Health Impact Assessment* found that zip codes in the Duwamish Valley were subject to substantially more health risks and had lower life expectancies than wealthier zip codes, and so the response has to do with outcomes: equalize those negative impacts across zip codes by any means necessary.

6 The EPA's final cleanup plan, *Record of Decision: Lower Duwamish Waterway Superfund Site* (Seattle: US Environmental Protection Agency, 2014), gives some of this history, as do various pages on the DRCC's website, www.duwamishcleanup.org.

7 The new mission statement is less focused on the cleanup and more on community empowerment. See www.duwamishcleanup.org/our-mission.
8 See Mark Purcell, *Recapturing Democracy: Neoliberalization and the Struggle for Alternative Urban Futures* (New York: Routledge, 2008).
9 Purcell, *Recapturing Democracy,* chap. 4.
10 EPA, *Record of Decision.*
11 William Daniell et al., *Health Impact Assessment: Proposed Cleanup Plan for the Lower Duwamish Waterway Superfund Site* (Seattle: School of Public Health, University of Washington, 2013).
12 Stefanie Spear, "Hip Hop Icon Macklemore Joins 'A River for All' Campaign," *EcoWatch,* July 7, 2014, www.ecowatch.com/hip-hop-icon-macklemore-joins-a-river-for-all-campaign-1881930029.html.
13 Duwamish River Cleanup Coalition, "Take a Look at Our Programs," accessed March 29, 2020, www.duwamishcleanup.org/about.
14 Duwamish River Cleanup Coalition, "Record of Decision: Final Cleanup Plan Community Fact Sheet" (Seattle: Duwamish River Cleanup Coalition/Technical Advisory Group, 2015).
15 See Duwamish River Cleanup Coalition, *Duwamish Valley Vision Map & Report* (Seattle: Duwamish River Cleanup Coalition/Technical Advisory Group, 2009).
16 Here they were only somewhat successful, gaining increased access only sometimes, and never to the point of being full partners in devising the draft plan. For more on the DRCC's efforts to democratize the Superfund process, see Purcell, *Recapturing Democracy,* chap. 4.
17 Duwamish Alive!, "Celebrate Earth Day with Duwamish Alive! April 20," *Westside Seattle,* April 11 2019.
18 See Duwamish River Cleanup Coalition, "Duwamish Valley Youth Corps Program," accessed March 29, 2020, www.duwamishcleanup.org/duwamish-valley-youth-corp-link.
19 The Port of Seattle estimates that originally the estuary "included approximately 1,450 acres of intertidal sand and mud substrate, nearly 1,300 acres of intertidal marsh, and approximately 1,450 acres of tidal swamp (or forested wetland)." Daniel Jack Chasan, "Can We Really 'Restore' the Duwamish, Seattle's Industrial River?" *Crosscut,* October 5, 2015, https://crosscut.com/2015/10/can-we-really-restore-seattles-industrial-river.
20 This is the sense in which they use the term *Duwamish Valley,* to capture the idea of the whole ecosystem in which the Duwamish sits. DRCC, "Duwamish Valley"; and Purcell, *Recapturing Democracy.*

CHAPTER 12

DRAWING THE THIN GREEN LINE

Throwing a Wrench in Carbon Commodity Chains

CORINA MCKENDRY AND NIK JANOS

> We can build part of a green wall on the West Coast by saying we aren't going to have these [fossil fuel] facilities in our city . . . it starts to have a profound effect that's far more than local.
>
> —CHARLIE HALES, MAYOR OF PORTLAND, OREGON, September 2014

A THIN GREEN LINE IS BEING DRAWN DOWN THE COAST OF Cascadia, one that is striving to interrupt the fossil fuel commodity chains that drive the global economy and are a major cause of climate change.[1] Local governments of this region, in partnership with social movements and tribes, are rejecting—and fighting beyond their borders to stop—the pipelines, railyards, and ports that are necessary to ship the great coal and oil reserves of Canada and the United States to international markets. In so doing, local governments are creating a new form of subnational climate politics, reshaping the carbonized territorial relationships of the national state and the fossil fuel–based global economy with potentially wide-reaching implications.

As they strive to intervene in fossil fuel commodity chains, subnational actors, particularly but not exclusively cities, are asserting what we call "embedded territorial sovereignty." We define embedded territorial sovereignty as the *collective decision-making mechanisms to establish rules for living together* that include but are not fully embodied by the nation-state and thus incorporate a multitude of actors and institutions operating at multiple scales and through a spectrum of adjacent and overlapping territories. Seen this way, the thin green line represents changing ideas and practices of territory, the state, and governance as people and institutions collide, on the one hand, to protect and expand the fossil fuel infrastructure that underlies carbon capitalism and, on the other, to change and ultimately to abolish fossil fuels as part of a vast, emerging project of decarbonization and democracy. As such, in these campaigns to disrupt oil, gas, and coal commodity chains, subnational actors are beginning to reshape climate politics and are creating new regional possibilities for envisioning a just, low-carbon future.

FOSSIL FUELS AND THE THIN GREEN LINE

The global economy runs on fossil fuels. Yet the sources of fossil fuels are geographically concentrated and by no means near where they will be consumed. This requires complex global commodity chains to facilitate the movement of fossil fuels from sites of extraction to sites of consumption. Crude oil is the most actively traded commodity in the world, and battles over control of oil fields have underpinned imperialism as well as liberation struggles.[2] Coal produces around 40 percent of the world's total electricity, and global coal exports doubled in the first decade of the twenty-first century.[3]

Oil and coal production and use have impacts that are both global and local, such as their most damaging unintended consequence, climate change. In addition, fossil fuels create high-paying jobs but also can destroy livelihoods through contaminating soil and water; they facilitate the health benefits that often accompany economic growth but also can cause severe health impacts to those directly exposed to their mining, transport, and combustion. These characteristics of fossil fuels—the distance between sites of production and sites of consumption and their simultaneous global and local impacts (both positive and negative)—create complex relationships between people, environments, and capital at various and intersecting scales.

The United States and Canada are both major coal and oil producers as well as consumers. Yet as coal use declines in North America and as increased oil production requires increased consumption, producers in both countries have been striving to increase exports, particularly to Asia, where economic growth is spurring demand for energy. The political problem facing both the Canadian and American fossil fuel industries' plans to increase exports, however, is that they have run straight into the green politics of Cascadia.

The major cities of Cascadia—cities with some of the strongest climate commitments and most organized environmental activism on the continent—stand between the huge fossil fuel reserves of North America and the growing international markets for these fuels. The idea behind the concept of the "thin green line" is that because of this geographic coincidence, the subnational governments of the region offer hope for restraining fossil fuel use and its associated carbon emissions by stopping the construction of the new oil and coal trains, ports, terminals, and pipelines that will be needed to get the fuels onto export-bound ships.[4] Indeed, evidence for this can be seen as cities and other subnational political actors in the region are intervening in the global commodity chains on which fossil fuels rely, thwarting the multibillion-dollar plans of some of the most powerful corporations in the world.

As addressed elsewhere in this book, the limitations and contradictions of Cascadia's green urban politics are significant. Yet the environmental politics of the region has the potential to throw a wrench in the carbon-based capitalist economy. To the extent efforts of the thin green line are successful and continue to expand, the region has a vital role to play in mitigating climate change by keeping North America's fossil fuels from being consumed.

DRAWING THE THIN GREEN LINE

Several examples illustrate the breadth of fossil fuel infrastructure projects that subnational governments have worked to stop across Cascadia. These efforts differ in their targets, tactics for preventing the project, and even the reason for their success or failure. Furthermore, though our analysis focuses on subnational governments, particularly cities, tribal governments have often been key players in these campaigns, and the violation of treaty rights or insufficient tribal consultation has been a major argument in the efforts to stop these infrastructure projects. This is vital to recognize as across North America Indigenous peoples are fighting to stop unwanted extractive infrastructure such as pipelines. We defer to the numerous Indigenous scholars and others working in this area to analyze these campaigns as parts of longer anticolonial struggles and as assertions of Indigenous sovereignty and tribal law that are disrupting settler colonial regulatory authority.[5] What this chapter hopes to add is an analysis of a different, and inchoate, assertion of sovereignty that is emerging from subnational governments. Local officials are insisting that fossil fuel infrastructure should not be built in or near their territories. We argue that as they do so, they are claiming sovereignty—understood here as the responsibility of ensuring well-being for their residents and the global community—as a defining feature of their efforts.[6]

One of the first instances of action along the thin green line occurred in Seattle between 2012 and 2015. The City of Seattle allied with the Lummi Tribe, citizens, and environmental activists to stop the construction of the Gateway Pacific Terminal at Cherry Point, Washington, which could have led to eighteen additional coal trains through Seattle per day.[7] City officials argued that the coal trains would have drastic consequences on residents' health, particularly the coal dust exposure in low-income neighborhoods, as well as negatively impact traffic, business, and industry along the rail route.[8] In addition, the Lummi Nation, whose reservation is adjacent to the proposed terminal, brought a lawsuit that argued the terminal would violate tribal rights because of the impact it would have on their fish harvest in Puget Sound. In 2015, the Army Corps of Engineers ruled on the side of the Lummi Nation, agreeing that significant evidence indicated that the export terminal would violate the Lummi fishing treaty rights.[9] The terminal project was dead.

In Vancouver, British Columbia, city officials attempted to intervene to stop federal approval of an expanded Trans Mountain Pipeline to carry oil from the Alberta tar sands to export facilities on the Burrard Inlet, on which Vancouver sits.[10] Though the pipeline route was not through city limits, Vancouver officials argued that the pipeline was inconsistent with broader provincial and municipal carbon reduction goals. They also raised the concern that the increased likelihood of an oil spill would damage the ecosystem of the bay and the city's valuable green brand.[11] Despite opposition from Vancouver, other area cities, seven First Nations, and, eventually, even the provincial government, in 2018, the federal government shocked many by spending $4.5 billion to buy the existing pipeline and unfinished expansion, as the owner had been scared off by the stiff political opposition.[12] In June 2019, the federal government officially approved the expansion. Though the Canadian Federal Court of Appeals had earlier asserted that the pipeline proposal failed to adequately consider environmental impacts and that consultation with First Nations had been inadequate, in early 2020, the Supreme Court refused to hear a case brought by Indigenous and environmental organizations, bringing the Trans Mountain Pipeline one step closer to construction.[13]

In 2016, Portland, Oregon, banned the expansion of all existing fossil fuel terminals in the city and any new terminals with a capacity of over two million gallons. Portland residents who worked to get the ban passed argued that it was necessary both because of the global climate impact of fossil fuels and because of the "enormous amount of health and safety risks" associated with fossil fuel storage and transportation.[14] Business groups including the Western States Petroleum Association fought the ordinance, claiming it violated the Commerce Clause of the US Constitution.[15] Though Portland's ban was reversed by the Oregon Land Use Board of Appeals, the case made its way to the Oregon Court of Appeals, which upheld the ordinance in January 2018.[16] Virtually all new fossil fuel infrastructure would henceforth be banned within the city's boundaries.

Also in 2016, the Vancouver, Washington City Council approved a ban on new oil refineries and facilities within the city. The facilities affected were actually fairly small, and the stated intention of the ban was to send a message to the Washington governor regarding a huge crude-by-rail facility that was being considered for the Port of Vancouver, over which the city did not have legal authority.[17] The large coalition working to stop the port expansion included residents and city officials in Vancouver, the Quinault Indian Nation, Ecojustice, the Columbia River Inter-Tribal Fish Commission, and others. In addition to strong concerns about the facility's climate impact, in a clear break with the supposed "jobs vs. the environment" argument, the local International Longshore and Warehouse Union also came out against the facility, citing the safety risk to its members of the increased oil trains and tankers.[18] In early 2018, the governor denied the necessary permit for the contested facility, stopping what would have been the largest port of its type in the nation.[19]

In 2019, the Jordan Cover Project, a proposed liquid natural gas export terminal, located in Coos County, Oregon, became another flashpoint of the thin green line. Canadian Pembina Pipeline Corporation first proposed the project in 2004, and it was granted initial approval by the US Department of Energy in 2014.[20] If constructed, the terminal will receive gas from a new Pacific Connector Gas Pipeline that will stretch 229 miles from a connection with existing pipelines near Malin, Oregon. The project has faced opposition since its inception, and in November 2019, environmental activists engaged in a sit-in at Oregon governor Kate Brown's office—twenty-one were arrested—to protest the terminal and demand that the governor reject the project.[21] Opponents argue that the project will violate tribal land, destroy old-growth forests, impact clean drinking water and clean water for salmon habitat, and contribute to rapid climate change.[22] In spring 2020, the US Federal Energy Regulatory Commission (FERC) voted two-to-one to give final approval to the project. Yet in a turn of events, shortly thereafter, Governor Brown came out against the Trump administration's fast final approval process, saying "I want to reiterate that I will not stand for any attempt to ignore Oregon's authority to protect public safety, health, and the environment."[23] Although officially approved by FERC, the project still has numerous regulatory gates to pass through, and state and regional activists and officials are working against it.[24] Its future is by no means secure.[25]

In each of these cases, elected officials, activists, and some local businesses and union leaders have articulated a wide variety of reasons for opposing new fossil fuel infrastructure. These efforts have been explicitly framed as global climate policies, as local governments and activists frequently insist that they must stop fossil fuel infrastructure in order to meet their commitments to global greenhouse gas reduction. Yet the actions also remain embedded within city-level concerns such as mitigating traffic and the health and safety impacts of the infrastructure on local communities. Many city officials have also highlighted the moral imperative of standing in solidarity with the treaty rights and sovereignty of Indigenous nations. In their efforts, subnational leaders assert that they have a legitimate stake in the global *and* the local impacts of fossil fuel transportation and use and a responsibility to take action to stop this infrastructure, whether or not it is within their official boundaries and legal jurisdiction. This phenomenon has immediate potential impacts on climate mitigation, as many of these efforts have contributed to stopping increased fossil fuel export capacity. Yet it also raises broader questions about the meanings of subnational sovereignty and territoriality in the time of climate change.

SPLINTERED SOVEREIGNTIES AND FLOWS OF CAPITAL

The politics of the thin green line illuminate a valuable contradiction in the role of cities in the globalized economy. Cities are seen as subservient to, or at least as tools of, flows of global capital. From resource extraction to the impact of corporate headquarters to investment capital's exacerbation of the housing crisis, flows of global capital have deeply shaped—and often scarred—Cascadia (see chapters 2, 9, and 10). Yet in the context of the thin green line, it is this very role of cities as conduits of carbon commodities that is enabling them to thwart these same processes of global flows.

Since the emergence of globalization in the 1990s, there has been a robust theorization of "global cities" as key nodes in facilitating the movement of global capital.[26] Very briefly, the 1980s saw a move away from efforts to promote and standardize national development toward policies promoting urban and regional competitiveness. Entrepreneurial cities became responsible for attracting increasingly mobile globalized capital, creating the "competition state."[27] Success gave certain cities power as sites where a great deal of the management of economic globalization unfolds.[28] These cities were connected to and shaped by, as well as helped to facilitate, the globalization of capital.[29]

Accompanying this process was "an increasing *splintering* of state regulatory activities and an intensified spatial *variegation* of state sovereignty at various spatial scales."[30] Under the competition state, rather than being defined by the physical control of territorial boundaries and what occurs inside them, sovereign "authority [became] vested in agents who *manage flows through space* or through action at a distance as much as in those who manage territories."[31] Though the primacy of nation-state sovereignty undoubtedly remains, this splintering has resulted in the rising importance of subnational actors clearly connected to processes of globalization but increasingly disarticulated from nation-state territory in their management of flows through space.[32] As Saskia Sassen asserts, these subnational sovereignties are not a move to local rule. Rather, the result has been "lumpy geographies of sovereign control."[33]

City interventions in fossil fuel commodity chains can be understood as part of this lumpy sovereign control of flows that occur locally but are connected to global agendas and are facilitated by the changing role of cities in the globalized economy. This is most basically because such efforts strive to contribute to global greenhouse gas reduction and more effective global governance of climate change. Yet the more common city climate efforts that focus on reducing a city's own emissions through efficiency, public transportation, renewable energy, and so forth can also be understood as contributing to wider processes of environmental governance.[34] What makes the efforts of the thin green line different is that cities, as well as other subnational governments, are challenging aspects of sovereignty that have long been the purview of states, particularly energy policy and large-scale, interregional infrastructure development. And to do so, they use their importance as conduits of capital flows as key points of intervention.

Fossil fuels are both a particularly abstract and a concrete commodity. On the one hand, the capital wrapped up in carbon commodity chains is very mobile, with fossil fuel companies accounting for 6 percent of global stock markets and crude oil being the most actively traded commodity in the world.[35] To the extent that global cities are major sites of virtual commodity trading (or, in the case of Seattle, a site of the invention of much of the computer technology that has enabled such trading), they facilitate and benefit from these capital flows.

On the other hand, the physical commodities of oil and coal are very much bound to physical territory and to movement across space.[36] This is a source of industry profits and political power but also fossil fuels' vulnerability. It is the inability of coal to be FedExed from Wyoming to China that makes Seattle's work to stop the expansion of the Cherry Point coal terminal so dangerous to the commodity chain. Likewise, if the necessary pipeline infrastructure is not built to get it to market, Alberta's bitumen is rendered worthless. In the usual story of cities and globalization, local governments are competing with one another to attract global capital and establish themselves as key nodes in global commodity chains. By contrast, in the interventions in global fossil fuel commodity chains discussed here, rather than facilitating the flows of the goods and capital that in this case literally drive the global economy (fossil fuels), cities of Cascadia are striving to break those flows.

The importance of capital movement in the globalized economy, and the accompanying "splintering" of sovereignty, has created spaces for cities and regions with the political will to do so to go against the preferences and policy goals of the federal governments of the United States and Canada, including the goal of increasing fossil fuel exports. This is why Cascadia's commodity chain interventions represent a potentially new and disruptive form of global climate politics. Though national states have facilitated the rescaling of urban governance and economic regulatory mechanisms to the subnational level for the purposes of the creation of the competition state, in so doing, they have also created opportunities for local and regional governments to engage in processes of territorialization that the architects of this devolution almost certainly did not have in mind. If regional and interurban competition is a defining feature of the globalized economy, when cities assert their sovereign authority through limiting commodity flows, the losers of this competition may in some instances be industries, such as fossil fuels, that elsewhere are vital elements of national economic strategy.

THE THIN GREEN LINE AS EMBEDDED TERRITORIAL SOVEREIGNTY

Globalization and the splintering of state regulatory activities have helped create the conditions of possibility in which cities of Cascadia are now reversing the competition state narrative in order to intervene in a vital global commodity chain. Yet in addition to presenting a new narrative on the role of cities in globalization, the thin green line also complicates what sovereignty and territoriality mean in this "lumpy" context of multiscalar climate governance.

Climate change challenges standard notions of sovereignty on the global scale. The standard understanding of sovereignty, wherein nation-states have complete control over everything within their clearly defined boundaries, has shaped global environmental politics since the inception of the environment as a recognized area of international concern in the early 1970s.[37] In climate politics, assertions of sovereignty have long been made by large industrializing nations such as China and India. They have argued that because their historical emissions were miniscule compared to that of the industrialized countries, any mandate to reduce emissions was an assault to their sovereign right to develop using efficient and inexpensive fossil fuels.[38] Ironically, as the United States has the largest historical responsibility for carbon emissions, President George W. Bush gave the same argument as his reason for not supporting the Kyoto Protocol, saying it unfairly impacted the United States' economy and interfered in the country's right to prioritize economic growth.[39] President Trump likewise asserted that the Paris Climate Agreement was a threat to US sovereignty.[40]

As the impacts of climate change have become increasingly urgent and visible, however, leaders of small island nations and other countries particularly vulnerable to climate change have been challenging the claims to a sovereign right to use fossil fuels made by China and the United States alike. Climate change, even more profoundly than other transboundary environmental issues, functions across political boundaries. Most notably, there is a deep disconnect between where greenhouse gases are emitted and where the most devastating effects of climate change will be, and are being, felt.[41] As leaders of highly vulnerable countries frequently assert in international forums and elsewhere, the fact that people in the places that have contributed the least to climate change are already being devastated by its impacts in ways that put their security, if not their nation's entire existence, in doubt shatters the illusion of territorial control that has been the promise of the system of sovereign states.[42]

If global climate injustice challenges theorizations of sovereignty as political control over territory vis-à-vis other sovereign nation-states, the thin green line offers an opportunity to re-envision sovereignty more broadly and richly as legitimate political authority.[43] From this perspective, sovereignty, rather than inherently oppressive and about centralizing control,[44] is instead "the institutions and practices . . . [that] have the unique power to establish binding rules for living together."[45] Despite the extent to which the existing infrastructure and processes of many states facilitate the worst treatment of the planet and of one another, sovereignty also provides the space through which human societies "can transform world-making—which is also collective self-making—into a deliberate act."[46] It becomes a process of collectively deciding what kinds of social and material infrastructure to make and, therefore, how we are to live together in the world.

In addition to creating institutions through which constructive world-making can be facilitated, this "legitimate authority" also holds the responsibility of securing the well-being of those within its jurisdiction. As climate change as well as the Trump administration's response to the COVID-19 outbreak painfully illustrate, the responsibility for securing human well-being, though often assumed to be vested in the nation-state, is too often abdicated by this particular political scale. Benjamin Barber asserts this unequivocally, arguing that national governments' failure to stop climate change and its accompanying threats to well-being "constitutes a default of sovereignty that both permits and demands action from alternative authoritative bodies: in our situation today, from regional and municipal authorities who assume responsibility for sustainability."[47] Cities, he continues, in their pursuit of efforts to mitigate climate change, are claiming the rights of sovereignty, particularly their responsibility to act to protect their residents when higher levels of government fail to do so.

Though Barber's analysis maintains a too-limited focus on localized greening efforts (white roofs, bike share, insulation, etc.), his argument that cities have a sovereign responsibility to protect residents is another way to understand the actions undertaken by the subnational governments constituting the thin green line. In resisting fossil fuel infrastructure, local leaders of Cascadia are insisting that doing so is vital for the social, and in many cases economic, well-being of their communities. Local leaders are also asserting that they have legitimate authority to intervene in national and international processes to secure this well-being in the face of federal government failure to reduce carbon emissions. This, we argue in agreement with Barber as well as with Jedediah Britton-Purdy's understanding of sovereignty as collective world-making, is fundamentally an assertion of sovereignty.[48]

In addition to being a declaration of subnational sovereign responsibility, the thin green line is also a new form of reterritorialization. By reterritorialization we mean the "reconfiguration and re-scaling of forms of territorial organization," such as between local governments and nation-states.[49] And it is in this reterritorialization that the lumpy sovereign control of globalization and the conceptualization of sovereignty as legitimate authority to make rules for living together merge. Sovereignty has traditionally been linked with territory. And like sovereignty, territory is often assumed to be relatively fixed, the clearly defined boundaries of a political space. Yet territory and the processes of territorialization that shape it are not as straightforward as they seem.

Territorial boundaries are the outcomes of historical, political, social, and economic actions and decisions.[50] The power of territory is therefore not the line on the map. Rather, its power rests in the political arrangements and contestations over the legitimate authority (sovereign ability) to measure and control space.[51] If understood as control over space, it is clear that territory is contested, fluid, and splintered. Borders of control are drawn and redrawn as the politics of the control of flows through space shift.[52]

Indeed, it is the contestation over flows through space that define territoriality, alongside the claims of legitimate political authority made by cities discussed above, that are defining the fights over carbon commodity chains and the thin green line. These actors are doing this in technical as well as legal aspects of the regulation of the flows of commodities through and within the specific territories of the nation-states in which they are embedded. Subnational actors are saying no to the nation-state and to private capital, thereby challenging both the material and immaterial control that federal governments and the fossil fuel industry have long had over the territories required to move fossil fuels.

As such, we see in the emergence of the thin green line what we are calling "embedded territorial sovereignty." We define this as the collective decision-making mechanisms to establish rules for living together that include but are not fully embodied by the nation-state and thus incorporate a multitude of actors and institutions operating at multiple scales and through a spectrum of adjacent and overlapping territories. This enables a challenge to the ways that struggles over power, rights, governance, and knowledge in the control of carbonized territorial relationships have long been naturalized.[53]

The naturalization of carbonized territorial relationships can be seen in the common assertion that calls for decarbonization are an attack on the American way of life, with "way of life" being "clearly defined as the unlimited and ever expanding ability of all American citizens to indulge in material consumption."[54] This makes calls for a shift away from a carbon economy ring as illegitimate, impractical, an attack on the middle class and even, in an inversion of calls for climate justice, a burden on the poor who live with the day-to-day difficulty of energy poverty.[55] The thin green line is challenging this naturalization of the centrality of fossil fuels to daily life.

RETHINKING CLIMATE GOVERNANCE

These emerging practices of embedded territorial sovereignty necessitate a rethinking and conceptual expansion of the role of cities in global climate governance. Specifically, the thin green line represents a new phase in urban climate change politics in North America. City and regional initiatives to prevent the export of fossil fuels demonstrate that subnational climate mitigation strategies are expanding beyond the still-important task of changing the built environment within the city limits toward an active pursuit of a wider transformation of the carbon-based global economy. Rather than, at best, connected islands of practice, municipalities, along with regional governments, social movements, and Indigenous peoples, are demonstrating what an extended urban politics of climate change might look like.

Further work needs to be done on how municipal, county, and provincial challenges to carbon commodity chains in North America link up with the efforts by non-state actors along other parts of these chains as well as political and macro factors in global demand that will play a major role in whether new nodes in carbon commodity chains are proposed and built. Also of particular importance will be understanding the relationship between subnational embedded territorial sovereignty as we have defined it here and the legal rights and sovereignties of Indigenous nations (see chapter 6). Yet what is already clear is that across the region, key political actors of Cascadia are challenging the assemblage of physical objects and flows through space that maintain the carbon-based economy and, as they do so, are extending the practice of governance in the age of climate change.

CONCLUSION

In this chapter, we have argued that the recent battles over carbon commodity chains that have flared across Cascadia represent a reconfiguration of the practice and understanding of sovereignty, territory, and the politics of climate change. The emergence of subnational entities declaring moral, legal, and legitimate authority that challenge the nation-state and capital implies claims of sovereignty, if we understand sovereignty as the exercise of legitimate political authority and the shaping of infrastructure, including territoriality. In claiming this sovereignty, the governments of Cascadia are shifting the territorial, scalar, and spatial configuration of the carbon commodity chains built throughout the twentieth century. We are thus seeing a reterritorialization of carbon commodity governance, an attempt to transform who and what is exercising control of the pathways of production and consumption.

Of course, power still matters, and for cities in particular, power remains circumscribed. In the examples offered here, city efforts to intercept fossil fuels relied on political and legal support from other sectors of the government, such as states, courts, or the Army Corps of Engineers. Clearly, "whether the repatterning of the scale of territories actually takes place in accordance with the desires and interests of a particular group of actors depends not only on the quality of the territorial proposals, but also crucially on the support and power of an interlocked multi-scalar coalition that provides the technical, scientific and discursive support to this reconfiguration."[56] In other words, and unsurprisingly, even as the thin green line challenges the territorial power of the fossil fuel industry and its allies in the federal governments of the United States and Canada, institutionalized political power remains vital.

Furthermore, whether the coalitions that have been built in each of these cases can be strengthened to continue to push for a revisioning of the relationship between Cascadia and fossil fuels remains to be seen. In the fights for the thin green line, city governments have stood in solidarity with tribal governments; labor unions have fought against expanding infrastructure in order to protect the health of workers and the communities through which the fossil fuels would pass; and environmental and social justice activists, as well as local officials, have articulated a multiscalar analysis of climate justice that demands an end to expanding fossil fuel infrastructure. In a region full of the contradictions of green capitalist urbanism, these coalitions hint at the potential for a shared struggle to create new forms of climate governance and more socially just ways to reimagine Cascadia in the face of climate change.

NOTES

The chapter's epigraph is quoted in Zahra Hirji, "Portland Bans New Fossil Fuel Infrastructure in Stand against Climate Change," Inside Climate News, December 15, 2016, https://insideclimatenews.org/news/14122016/portland-oregon-ban-fossil-fuels-oil-and-gas-pipelines-coal-global-warming.

1 Though the term *thin green line* has been used in a wide variety of capacities, we would like to acknowledge the Seattle-based think tank the Sightline Institute for our inspiration regarding the use of this term. Neither author has any affiliation with the Sightline Institute. See Sightline Institute, "The Thin Green Line," accessed May 16, 2018, www.sightline.org/research/thin-green-line/.

2 Will Hall-Smith, "Top 10 Most Traded Commodities," IG Analyst, September 5, 2018, www.ig.com/en-ch/trading-strategies/top-10-most-traded-commodities-180905; and Timothy Mitchell, "Carbon Democracy," *Economy and Society* 38, no. 3 (2009): 399–432.

3 "Electricity Production from Coal Sources (% of total)," World Bank, accessed April 19, 2018, https://data.worldbank.org/indicator/EG.ELC.COAL.ZS; and "Coal Trade around the World," CarbonBrief, May 3, 2016, www.carbonbrief.org/mapped-the-global-coal-trade.

4 Sightline Institute, "Thin Green Line."

5 For example, Kyle Powys Whyte, "The Dakota Access Pipeline, Environmental Injustice, and U.S. Colonialism," *Red Ink* 19, no. 1 (2017): 154–69; Tyler McCreary and Jerome Turner, "The Contested Scales of Indigenous and Settler Jurisdiction: Unist'ot'en Struggles with Canadian Pipeline Governance," *Studies in Political Economy* 99, no. 3 (2018): 223–45; Gordon Christie, "Indigenous Authority, Canadian Law, and Pipeline Proposals," *Journal of Environmental Law and Practice* 25 (2013): 189–215; and Dina Gilio-Whitaker, *As Long As Grass Grows: The Indigenous Fight for Environmental Justice, from Colonization to Standing Rock* (Boston: Beacon Press, 2019).

6 Benjamin R. Barber, *Cool Cities: Urban Sovereignty and the Fix for Global Warming* (New Haven, CT: Yale University Press, 2017).

7 Mike Lindblom, "Report: Coal Trains Planned for State Carry Issues, Bring Jobs," *Seattle Times*, July 25, 2014.

8 Parametrix, "Coal Train Traffic Impact Study," October 2012, www.seattle.gov/Documents/Departments/SDOT/About/DocumentLibrary/Reports/ 121105PR-CoalTrainTrafficImpactStudy.pdf; and Richard Conlin, "Stopping the Coal Trains," October 15, 2013, http://conlin.seattle.gov/2013/10/15/stopping-the-coal-trains/.

9 US Army Corp of Engineers, "Army Corps Halts Gateway Pacific Terminal Permitting Process," May 9, 2016, www.nws.usace.army.mil/Media/NewsReleases/Article/754951/army-corps-halts-gateway-pacific-terminal-permitting-process/.

10 David P. Ball, "'No Rationale Given,' Say Rejected Kinder Morgan Intervenors," The Tyee, April 4, 2014, http://thetyee.ca/News/2014/04/04/Rejected-Kinder-Morgan-Intervenors/.

11 City of Vancouver, *Greenest City 2020 Action Plan: 2014–2015 Implementation Update* (Vancouver: City of Vancouver, 2015).

12 Laura Kane, "Federal Court Allows Half of Trans Mountain Pipeline Legal Challenges to Proceed," *Global News*, September 4, 2019, https://globalnews.ca/news/5855002/trans-mountain-pipeline-court-appeal/.

13 Alastair Spriggs, "Court Overturns Kinder Morgan Pipeline Approval; Project Future Questioned," The Tyee, August 30, 2018, https://thetyee.ca/News/2018/08/30/Court-Appeal-Overturns-Kinder-Morgan-Pipeline-Approval/; and Rhianna Schmunk, "Supreme Court of Canada Will Not Hear B.C. Groups' Challenges against Trans Mountain Pipeline Expansion," *CBC News*, March 5, 2020, www.cbc.ca/news/canada/british-columbia/trans-mountain-pipeline-appeals-supreme-court-of-canada-1.5486592.

14 Quoted in Bruce Lieberman, "What Other Cities Can Learn from Portland's Fossil-Fuel Ban," Yale Climate Connections, April 25, 2018, www.yaleclimateconnections.org/2018/04/portlands-fossil-fuel-ban/.

15 Ted Sickinger, "Portland's Fossil Fuel Infrastructure Limits Overturned," *Oregonian/OregonLive*, updated January 9, 2019, www.oregonlive.com/business/2017/07/land_use_board_strikes_down_po.html.

16 Pete Danko, "Portland Fossil-Fuel Infrastructure Ban Survives Court Challenge," *Portland Busines Journal*, August 1, 2018, www.bizjournals.com/portland/news/2018/08/01/portland-fossil-fuel-infrastructure-ban-survives.html; and Kevon Paynter, "First U.S. City to Ban Fossil Fuel Expansion Offers Roadmap for Others," *Yes! Magazine*, February 5, 2018, www.yesmagazine.org/environment/2018/02/05/first-us-city-to-ban-fossil-fuel-expansion-offers-roadmap-for-others/.

17 Lauren Dake, "Unaminous Vote Bans Oil Facilities in Vancouver," *The Columbian*, July 18, 2016, www.columbian.com/news/2016/jul/18/unanimous-city-council-vote-bans-oil-facilities-in-vancouver/.

18 Molly Solomon, "How a Grassroots Effort in Vancouver Fought Big Oil—and Won," Oregon Public Broadcasting, February 13, 2018, www.opb.org/news/article/vancouver-washington-oil-terminal-grassroots-community-oil-companies/.

19 Rick Anderson, "How Forces Combined again in Washington State to Reject Yet Another Oil Terminal," *Los Angeles Times*, February 4, 2018.

20 "Coos Bay LNG Pipeline and Terminal Project Gets Authorization," Wild Coast Compass, July 8, 2020, www.wildcoastcompass.com/coos-bay-lpg-pipeline-and-terminal-project-gets-authorization/. For company information, see "The Jordan Cove Project: Safe. Clean. Reliable," Jordan Cove LNG: A Pembina Company, accessed November 28, 2020, www.jordancovelng.com/project.

21 Lee van der Voo, "It's His Land. Now a Canadian Company Gets to Take It," ProPublica, October 1, 2020, www.propublica.org/article/its-his-land-now-a-canadian-company-gets-to-take-it; and "21 Protesters Arrested after Taking over Oregon Governor's Office to Oppose Jordan Cove Project," Oregon Public Broadcasting, November 21, 2019, www.opb.org/news/article/jordan-cove-sit-in-kate-brown-office/.

22 Andy Rowell, "Protesters Demand That Oregon Gov. Brown #StopJordanCove," Oil Change International, November 22, 2019, http://priceofoil.org/2019/11/22/protesters-demand-that-oregon-gov-brown-stopjordancove/.

23 Ted Sickinger, "Federal Regulators Uphold Approval of Jordan Cove LNG Project in Coos Bay, Teeing Up Potential State Lawsuit," *The Oregonian*, May 22, 2020, www.oregonlive.com/politics/2020/05/federal-regulators-uphold-approval-of-jordan-cove-lng-project-in-coos-bay-teeing-up-potential-state-lawsuit.html.

24 Alex Schwartz, "Despite Federal Export Approval, Jordan Cove LNG Still in Uphill Battle," *Herald and News* (Klamath Falls, OR), July 8, 2020, www.heraldandnews.com/news/local_news/despite-federal-export-approval-jordan-cove-lng-still-in-uphill-battle/article_1b471c90-ae78-533b-936f-a226e97fd309.html.

25 At the time of this writing, it is an open question how ambitious on climate the Biden administration will be, though the strong actions taken in the first few weeks of the new administration are promising.

26 John Friedmann, "The World City Hypothesis," *Development and Change* 17, no. 1 (1986): 69–83; Darel E. Paul, "Re-scaling IPE: Subnational States and the Regulation of the Global Political Economy," *Review of International Political Economy* 9, no. 3 (2002): 465–489; Saskia Sassen, *The Global City: New York, London, Tokyo* (Princeton, NJ: Princeton University Press, 2001); and Saskia Sassen, "Locating Cities on Global Circuits," *Environment and Urbanization* 14, no. 1 (2002): 13–30.

27 David Harvey, "From Managerialism to Entrepreneurialism: The Transformation in Urban Governance in Late Capitalism," *Human Geography* 71 (1989): 3–17; and Neil Brenner and David Wachsmuth, "Territorial Competitiveness: Lineages, Practices, Ideologies," in *Planning Ideas That Matter: Livability, Territoriality, Governance, and Reflective Practice*, ed. Bishwapriya Sanyal, Lawrence J. Vale, and Christina D. Rosan (Cambridge, MA: MIT Press, 2012), 179–206.

28 Sassen, "Locating Cities."

29 Manuel Castells, *The Informational City: Information Technology, Economic Restructuring, and the Urban-Regional Process* (Oxford: Basil Blackwell, 1989); and Neil Brenner, "Globalisation as Reterritorialisation: The Re-scaling of Urban Governance in the European Union," *Urban Studies* 36, no. 3 (1999): 431–51.

30 Neil Brenner, "Urban Locational Policies and the Geographies of Post-Keynesian Statehood in Western Europe," in *Cities and Sovereignty: Identity Politics in Urban Spaces*, ed. Diane E. Davis and Nora Libertun de Duren (Bloomington: Indiana University Press, 2011), 164–65, emphasis in original.

31 John Agnew, "Sovereignty Regimes: Territoriality and State Authority in Contemporary World Politics," *Annals of the Association of American Geographers* 95, no. 2 (2005): 442, emphasis added.

32 Nils Gilman, "The Twin Insurgency," *American Interest* 9, no. 6 (2014): 7–14; Saskia Sassen, *Territory, Authority, Rights: From Medieval to Global Assemblages* (Princeton, NJ: Princeton University Press, 2006); and Saskia Sassen, "When Territory Deborders Territoriality," *Territory, Politics, Governance* 1, no. 1 (2013): 21–45.

33 Sassen, *Territory, Authority, Rights*, 54–55.

34 Corina McKendry, *Greening Post-Industrial Cities: Growth, Equity, and Environmental Governance* (New York: Routledge, 2018); and Craig Johnson, Noah Toly, and Heike Schroeder, eds., *The Urban Climate Challenge: Rethinking the Role of Cities in the Global Climate Regime* (London: Routledge, 2015).

35 Matt McGrath, "Carbon 'Bubble' Could Cost Global Economy Trillions," *BBC News*, June 4, 2018, www.bbc.com/news/science-environment-44357243.

36 Trafigura, *Commodities Demystified: A Guide to Trading and the Global Supply Chain*, 2nd ed. (Geneva: Trafigura, 2019), www.commoditiesdemystified.info/pdf/CommoditiesDemystified-en.pdf#Commodities-Demystified.

37 Steven Bernstein, *The Compromise of Liberal Environmentalism* (New York: Columbia University Press, 2001); see also Joao Augusto de Araujo Castro, "Environment and Development: The Case of the Developing Countries," in *World Eco-Crisis: International Organizations in Response*, ed. David A. Kay and Eugene B. Skolnikoff (Madison: University of Wisconsin Press, 1972).

38 Gørild Heggelund, "China's Climate Change Policy: Domestic and International Developments," *Asian Perspective* 31, no. 2 (2007): 155–91.

39 Associated Press, "Bush: Kyoto Treaty Would Have Hurt Economy," NBCNews.com, June 30, 2005, www.nbcnews.com/id/8422343/ns/politics/t/bush-kyoto-treaty-would-have-hurt-economy/#.V0yg8ZMrJDA.

40 Donald Trump, "Statement by President Trump on the Paris Climate Accord," Washington, DC, June 1, 2017, https://trumpwhitehouse.archives.gov/briefings-statements/statement-president-trump-paris-climate-accord/.

41 Anil Agarwal, Sunita Narain, and Anju Sharma, "The Global Commons and Environmental Justice—Climate Change," in *Environmental Justice: Discourses in International Political Economy*, ed. John Byrne, Leigh Glover, and Cecilia Martinez (New Brunswick, NJ: Transaction Publishers, 2002), 171–99; and Chukwumerije Okereke, "Climate Justice and the International Regime," *WIREs Climate Change* 1, no. 3 (2010): 462–74.

42 Darren Hoad, "The 2015 Paris Climate Agreement: Outcomes and Their Impacts on Small Island States," *Island Studies Journal* 11, no. 1 (2016): 315–20.

43 Agnew, "Sovereignty Regimes"; and Barber, *Cool Cities*.

44 See, for example, Geoff Mann and Joel Wainwright, *Climate Leviathan: A Political Theory of Our Planetary Future* (London: Verso, 2018).

45 Jedediah Britton-Purdy, "The World We've Built," *Dissent*, July 3, 2018, www.dissentmagazine.org/online_articles/world-we-built-sovereign-nature-infrastructure-leviathan.

46 Britton-Purdy, "The World We've Built."

47 Barber, *Cool Cities*, 19.

48 Britton-Purdy, "World We've Built."

49 Brenner, "Globalisation as Reterritorialisation," 433.

50 Rutgerd Boelens et al., "Hydrosocial Territories: A Political Ecology Perspective," *Water International* 41, no. 1 (2016): 1–14.

51 Stuart Elden, "Land, Terrain, Territory," *Progress in Human Geography* 34, no. 6 (2010): 799–817.

52 Agnew, "Sovereignty Regimes"; Elden, "Land"; and Nik Janos, "Urbanising Territory: The Contradictions of Eco-cityism at the Industrial Margins, Duwamish River, Seattle," *Urban Studies* 57, no. 11 (2020): 2282–99.

53 See Boelens et al., "Hydrosocial Territories," for a theorization of this idea in the context of water.

54 Jean-Daniel Collomb, "The Ideology of Climate Change Denial in the United States," *European Journal of American Studies* 9, no. 1 (2014): 9.

55 Collomb, "Ideology"; and Jen Schneider et al., *Under Pressure: Coal Industry Rhetoric and Neoliberalism* (London: Palgrave Macmillan, 2016).

56 Boelens et al., "Hydrosocial Territories," 5.

CONCLUSION

NIK JANOS AND CORINA MCKENDRY

THE AUTHORS IN THIS BOOK ENDEAVOR TO UNDERSTAND and analyze what we call *urban Cascadia*: not just the major cities of the region but the pathways, circuits, networks, socioecologies, built environments, and nonhuman animals and plants that are interconnected in an increasingly urbanized bioregion, which is also intersected and shaped by national and planetary processes and institutions. In particular, the book uses the Cascadia imaginary—both its left bioregional and eco-capitalist variants—as a reference point to examine the three core historical and contemporary developments at play in the region: capitalist urbanization of nature, stretching from European colonialism to contemporary green urbanism; the creation of social and environmental inequalities that contradict the green, progressive image of the region; and the movements and efforts to construct social and environmental justice out of the first two. Though we recognize that these are not unique to Cascadia, the strength of the green imaginary in the region makes them particularly pronounced.

In this conclusion, using the insights from this book's authors, we reflect on the ongoing challenges and contradictions faced by the people who inhabit Cascadia. We have separated the three challenges and contradictions in this way: 1) infrastructure and environmental contradictions, 2) Indigenous peoples and settler urbanization, and 3) democracy and the green imaginary. We will reflect on how these challenges and contradictions help us better understand the core questions that motivated this book: What does the urban environmental history of the region tell us about the processes and patterns of historical and contemporary capitalist urbanization? What steps are being taken by Indigenous and non-Indigenous people to construct genuine politics of recognition, reconciliation, and Indigenous sovereignty throughout the region? Moving beyond the green imaginary, what ways are people in Cascadia addressing social and environmental inequities, environmental degradation, resource use, and climate change and decarbonization, and what are the barriers faced in achieving a more democratic, just, and sustainable region? And what can in-depth research and thinking about Cascadia tell us about environmental and urban politics, environmental inequalities, the relationship between cities and rural areas, and the politics of environmental change more broadly? At the end of this conclusion, we briefly consider the future of the green imaginary of Cascadia.

INFRASTRUCTURE AND ENVIRONMENTAL CONTRADICTIONS

Diverse people, societies, nation-states, and regions are facing ongoing local, regional, and planetary environmental crises, such as water and air pollution, resource extraction and availability, disease, and global climate change. Some of these are emergent, while others are rooted in colonial and capitalist economies of the past several hundred years. As many of the authors in this book have shown, the burdens and benefits of these systems and their associated crises are unevenly distributed along the lines and intersections of Indigenous vs. non-Indigenous, race and ethnicity, gender, and class.

To the extent that "greenness" is one of the defining features of the Cascadian imaginary, the environmental destructiveness at the heart of the settler colonial urbanization project is a core contradiction of the region. Yet it is also one of hybridity, in which the socionatures and green politics of the region are not simply branding (though they certainly are that as well) but represent complex relationships between humans and nonhuman nature, tied together by nuanced and often contradictory engagements with global and regional infrastructure.

In chapter 3, Thaïsa Way and Ken Yocom illustrate this complexity through the concept of "infrastructural wilderness." This concept allows us to analyze how city making, urbanization, and nature are intertwined in the process of bringing the urban and the wild into a coproducing relationship. Their history reveals that urban Cascadia is not simply made in the cities; rather, it is a historical process of extending urbanization from hinterland to urban concentration. By decentering the city, their chapter considers how so-called spaces of wild nature are constituted as central parts of urban nature by the linkages of infrastructure that binds "wilderness" and "urban" together. Similarly, Alida Cantor and Alexander Reid Ross help to move the definition of urbanization away from the central city to a definition of extended urbanization that connects rural, exurban, and urban areas into new political and ecological relations. Using the politics of water, their chapter shows how new tensions and contradictions embedded in the governance of water and urbanization are leading to new exurban political ecologies. Both chapters open space to think through how infrastructure enables urban linkages with nature outside the city.

With an even more expansive examination of regional infrastructure, in chapter 2, Nik Janos traces the urbanization of Cascadia through the processes that transform nature into energy, transportation, and water. Like Way and Yocom and Cantor and Ross, Janos shows that these historical and ongoing extended socionatural transformations are connected to the processes of city and suburb building. He also illustrates the deep connection between infrastructure and capital: for example, the hydroelectricity from Cascadia's dams powers both personal computers and the construction of jumbo jets. Urban Cascadia has been in the process of becoming in a material and regulatory sense for the better part of a hundred years. Yet the boosters and managers of eco-capitalist Cascadia are pushing hard to further integrate the region politically and economically, with infrastructure such as a new high-speed rail, like the dammed rivers before it, key to these connections.

The infrastructure of the region has transformed socionature in many destructive ways, from deforestation to dams destroying salmon runs. Yet we show another side of this relationship in the book's final chapter. We demonstrate that the importance of infrastructure within Cascadia, in this case the region's role as a site of export-oriented oil, gas, and coal infrastructure, is opening up new spaces of resistance to global corporations and new tactics of subnational climate governance. Even as eco-capitalist Cascadian leaders are placing their climate efforts concretely within the green brands they have fostered, through building the "thin green line," they are also resisting climate injustices and creating new forms of urban territoriality and sovereignty that are reshaping the politics of decarbonization.

Together, these chapters illustrate the ways that energy, infrastructure, and territory are key terrains for producing—and for addressing—urban and environmental crises. They highlight how collaboration and conflict reside alongside these hybrid socionatures and the infrastructure that has created them. The larger lesson from these chapters is that the socionature that we have might not be ideal, and it might not be just, but it is the socionature that we have to work with and out of which any more socially just and environmentally sustainable future must be created.

INDIGENOUS PEOPLES AND SETTLER URBANIZATION

A foundational challenge and contradiction of Cascadia is the historical and ongoing settler colonial dispossession of Indigenous lands and ways of life. In his chapter, Coll Thrush recasts urban history as Indigenous history. He shows that the founding of the big cities of Cascadia was deeply embedded within the process of settler colonial displacement and dispossession. However, rather than shrinking to the rural margins, Indigenous people were integral to the making of these cities. Settler colonialism in Cascadia, therefore, was not only a project of the empires and nation-states of Europe, the United States, and Canada. It was also a project of local, urban colonialism in Seattle, Vancouver, Victoria, and Portland.

Yet the realities of settler colonialism and the struggles of Indigenous peoples for sovereignty and survival are not consigned to the past. Thrush emphasizes the agency of Indigenous people with the concept of Indigenous survivance. The concept of survivance recognizes the continued Indigenous resistance, adaptation, and, in some cases, flourishing in the face of centuries of settler colonialism. Survivance is clearly illustrated in Danica Miller's chapter about the Puyallup Tribe's fight for sovereignty and the right to harvest salmon, from the earliest treaties with the United States through the Fish Wars in 1960s and 1970s. Miller gives a powerful account of the Puyallup's cultural and religious meanings and practices of salmon fishing and the intense battles over the centuries to defend Indigenous sovereignty and wrestle control of salmon and socioecological spaces away from the colonial state. As these chapters illustrate, the Indigenous peoples of Cascadia continue to shape the region.

Jeffrey Sanders explores another contradiction at the center of urban Cascadia connected to the built environment and political and legal structures of settler colonialism. Stretching from the time of the settler colonialists to the internet capitalists of today, the region has been imagined as a progressive utopia, one that blends nature worship with technological worship. Yet this utopia has been, and continues to be, built on the dispossession and exploitation of Indigenous people, the working class, and the nonhuman world. In his chapter, Sanders provides an unsettling exposé that the yearnings of utopian Cascadia, whether countercultural or capitalist, have deep roots in colonial and industrial capitalist urbanization imaginaries and built environments.

ENVIRONMENTAL INJUSTICE AND THE GREEN IMAGINARY

One of the central motivations of this book was to analyze the tension between the image, branding, and discourse of green, progressive Cascadia on the one hand and the material and structural socioenvironmental inequalities on the other. As discussed above, Sanders historicizes the "ecotopia" and "techno-utopia" underpinnings of urban Cascadia and shows how these two visions are deeply interwoven in the imaginaries of eco-capitalist and left bioregionalism. Multiple chapters in the book explore the contradiction between these green utopian imaginaries and the actually existing social and environmental inequalities of the region. For example, Erin Goodling and Mark Purcell explore the politics of Superfund pollution cleanup in Portland and Seattle, respectively. Goodling illuminates the contradiction between the environmental rhetoric and imaginary of Portland, and Cascadia more generally, and historical and existing environmental injustices. People of color, immigrants, Indigenous people, and the working class have been disproportionately exposed to pollution and urban environmental inequalities throughout the history of Portland. Her "peoples' history" illustrates how frontline communities are shaping the politics of river cleanup in ways that center injustices and complex relationships with place while challenging official proclamations of the greenness of Portland.

Similarly, Purcell's chapter highlights the history of urban environmental inequalities along the Duwamish River in Seattle. Purcell uses the Duwamish River Cleanup Coalition and the politics of pollution cleanup as a lens to explore the relationship between "justice" and "democracy." He argues that too often social movements favor just outcomes to democracy, the latter understood as a collective endeavor to manage our common affairs. He asks us to reframe our conversation from justice to democracy and, in doing so, opens up new questions about how urban nature is produced and who and what constitutes a political community. Provocatively, he also asks us to consider ecological democracy in which humans and nonhumans are joined in a political community that requires listening, understanding, and notions of power through and beyond human decision-making. Goodling and Purcell help us see that democracy has the potential to shape the process by which just outcomes can be achieved; the values of the green imaginary of urban Cascadia; and the meanings of rivers, ecosystems, and nonhuman animals, as well as the place of humans alongside them.

Another source of environmental injustice is the inequality produced through capitalist urban (re)development and the processes of gentrification. Jennifer Rice and Giuseppe Tolfo explore these processes and how they form the dark side of the green, livable city. Rice provides an examination of how urban politics and inequalities emerging within Cascadia, in this case Seattle, are being extended across space and location via Amazon. Her interlocuters give this phenomenon the name "Seattle-ization," referencing social dislocation and inequality rather than the progressive and green urbanism that the cities in the region are noted for. This chapter also amplifies the voices of activist groups and residents who are pushing back against the power of Amazon, and Seattle-ization more broadly, in order to create a more just region. Tolfo, in his chapter on Vancouver, takes a historical approach to the relationship between urban redevelopment and the making of inequalities. Recognizing that the city's discourses of livability and sustainability have often benefited the middle-class and well-to-do, he argues that emerging politics of recognition and reconciliation will not go far enough if not coupled with economic reparations for Indigenous people and people of color, as well as Indigenous rights of self-determination.

Looking at the region's housing crisis from a different perspective, Corina McKendry raises important questions about the form of "sustainability" and "green urbanism," such as the shape, size, physical quality and quantity, and location of urban development. At heart these are not simply about aesthetics but about how people experience and perceive the spaces of everyday life, deeply imbued with values as well as practicalities. Moreover, the chapter highlights the debates and conflicts over the processes of urban governance and green urbanism under capitalism. McKendry indicates that the sticking point is not sustainability per se but rather questions of who wins or loses, who participates and who does not, and how fast or slow change should take place. On this last point, McKendry helps us understand that the temporal aspect of social and urban transformation is vitally important to the governance and politics of environmental change. The tension between the speed of change—at any scale—and democracy, deliberation, and lived experience is one that needs further exploration in a time of accelerating environmental crises.

THE FUTURE OF CASCADIA

As discussed in the introduction to this book, Cascadia is not a coherent political territory or a place on any official map. Rather, Cascadia is a political imaginary, one that is emergent and contested. Two mainstream imaginaries have emerged over the past fifty years, what we call left bioregional and eco-capitalist Cascadia. The former is rooted in the bioregionalism and localist movements of the 1970s and is predicated on a political community that emerges from the regional biome. It is a place-based inclusive, democratic, localist, and bottom-up idea of politics. Left bioregionalists are critical of both global capitalism and the territoriality of nation-states that divide people on political boundaries. However, as we wrote in the introduction, bioregionalism also has its white supremacist and neo-Nazi wing that draw, uncomfortably, on many of the same values as left bioregionalists but accompanied by explicit desires for racial exclusion.

Eco-capitalist Cascadia also takes notes from bioregionalism yet is a top-down idea of politics predicated on local governmental and economic elites constructing interwoven economies and governance structures in the region. Eschewing to some degree typical nation-state territoriality, proponents of eco-capitalist Cascadia adhere to the power of a nested regionalism to bolster locally sited yet global corporations and job markets within the pressures of competitive globalization. There are, of course, overlaps between these two ideologies. However, it is an open question whether Cascadia will emerge as either a coherent political and economic territory, a unifying collective identity, or as a broad-based and inclusive social movement. We thus reflect on the future of Cascadia by pointing to possibilities and limits of regionalism.

WHAT IS THE PROSPECT OF REGIONAL IDENTITY AND GOVERNANCE?

The prospects of a regional identity or a co-governed economic and political territory called Cascadia hinge on how people define regionalism in the first place. As discussed, proponents and skeptics might argue whether Cascadia is an actually existing region. Though this is not the place to review the literature on urban regionalism, it is worth noting that urban scholars have been keen on understanding the proliferation of urban and metropolitan regionalism since the 1990s.[1] In particular, scholars have argued that globalization and reterritorialization have led to the emergence and rise of global cities and the expansion of urbanized megaregions in various parts of the world.[2] To what extent is Cascadia an urbanizing region, and to what extent does the political imaginary precede or succeed economic and political regionalism? To answer these questions, it is instructive to return to Smith's three definitions of Cascadia as physical, functional, and ideological.[3] We can use each definition to gain clarity on the possibilities and limits of Cascadia.

The first and foundational definition of Cascadia is the physical definition of a specific and unique biome of North America (see map I.1 in introduction). All other definitions of Cascadia more or less rest on whether one accepts that there is in fact a distinct biome that can be measured empirically and can be tied to specific political and social claims. However, there is reason to be skeptical of the bioregionalist idea that if there is a distinct and coherent biome, then politics naturally flows from it.

As this book has illustrated, social and ecological conditions are coproduced through the forces of capitalism, states, and social participation in uneven and unequal ways. Sanders shows how the colonial projects in what became British Columbia, Washington, and Oregon were predicated on idealized notions of wild and pristine nature. These new territories and spaces were used to construct an ecotopia, a green colonial project that was to stand in contrast to eastern North America, where colonists felled every tree and mined every mountain. Moreover, Thrush and Miller show how, at its heart, the colonial project was founded on the dispossession of Indigenous ways of life, the transfer of sovereignty, the appropriation of nature and land, and ultimately the production of a new socionature people now call Cascadia. Way and Yocom further illustrate the ways that even the wildest, most "natural" parts of the region are deeply embedded in—and have been transformed by—processes of capitalist urbanization. That Cascadia as a biome is so clearly inseparable from this ongoing history should make us question the claim that nature is the wellspring of politics.

The second definition of Cascadia is functional, dealing with the infrastructural, territorial, political, and economic roles Cascadia plays in an interconnected regional, national, and international space. Evidence appears to show that there is indeed an emerging functional region defined by transportation, energy, and water infrastructure and by the connected urban, suburban, exurban, and rural spaces. In some ways, urban Cascadia is analogous to other urbanizing megaregions around the world, such as Boston–New York–Washington, DC; the Øresund Region linking Zealand, Denmark, and Skåne Sweden; or the Pearl River Delta Metropolitan Region in southern China. Of course, important differences with regard to economic intensity, density, and governance practices, as well as lived experience, in each region make direct comparisons tricky. Nonetheless, as Janos, Way and Yocom, and Cantor and Ross illustrate in their chapters and as we show in the final chapter, there is analytical usefulness of thinking about the ways that Cascadia does operate as a functional region.

Beyond its analytical usefulness, the functional definition is appealing to proponents of eco-capitalist Cascadia, those who are actually building and governing these large extended urban infrastructures and amalgamations of people and built environments. Critical analysis is warranted of these often top-down, elite-driven forms of urbanization. Many of the chapters of this book explore how capitalist urbanization and city, provincial, and national governments produce deep inequalities and uneven socionatures throughout the region. Rice and Tolfo, for example, show how urban displacement is tied not only to new forms of urban development but also to older forms of settler urbanism. White and wealthier people and institutions reap most of the benefits and rewards, whereas the negative consequences of urban Cascadia are borne by people of color, Indigenous people, and the working class. These material inequalities form one-half of the dyad of the green contradiction. The other half is the idealized green ideology of Cascadia that we turn to next.

The third definition of Cascadia relates to how it is constructed as a political ideology rooted in naturalism and regionalism. As discussed in the introduction, proponents of the Cascadia ideology, of all political stripes, use it to create a shared sense of belonging, culture, history, and community, often deploying flags and other branding elements. However, as we noted, there are political ideological differences between these imaginaries. On the one hand, the left bioregionalists anchor their version of Cascadia in the land and, at their best, in acknowledging the historical injustices of settler colonialism. They also craft their imaginary as an alternative to capitalist globalization and the nation-state, which they see as the prime drivers of environmental destruction, social fragmentation, and exploitation. On the other hand, the eco-capitalists borrow the bioregionalism of the eco-left but construct their version of Cascadia as an integrated and accelerated regional political economy that stands in contrast to and in competition with other political scales and economic blocs across the planet.[4] Yet both of these ideologies of Cascadia share three things in common: the naturalization of politics, the supposition of a *sui generis* political community, and a post-political obfuscation of the tensions and contradictions inherent in their vision.

It is on these points that skeptics and critics of both eco-left and eco-capitalist ideologies of Cascadia might rightly point out a central contradiction: the question of who belongs and who the legitimate political actors are in this supposedly naturally defined community. The history of environmentalism, including the naturalized mythos and politics of Cascadia, is unfortunately replete with stories of the way nature is used to draw lines between worthy and unworthy political subjects.[5] Left bioregionalism is an ideology of solidarity, community, and autonomy. Yet it cannot escape the twin phenomena of inequalities in the construction of ideology (who are the doers) and inequalities in outcomes of ideology (who belongs).

However, it is eco-capitalist Cascadia that most clearly embodies this contradiction. This is because their version of Cascadia is decidedly top-down, elitist, and most clearly uses Cascadia as a post-political brand of liberal environmentalism that has become *de rigueur* of regional political and economic elites. This is not to say there is nothing profound or transformational about their version of Cascadia, especially as it stands in contrast to political and economic elites in other parts of North America or at other political scales. Yet as many of the chapters of this book clarify, the green capitalist imaginary rarely withstands scrutiny when evidence of displacement, gentrification, health disparities, environmental problems, and the ongoing struggles of Indigenous nations are stacked against it. Whether left bioregional or eco-capitalist, Cascadia remains an unsettled and contentious imaginary.

TOWARD DEMOCRACY AND JUSTICE

So what is the future of Cascadia? Will it remain a metaphorical and ideological reference point, always on the horizon of becoming? Will it emerge as a contested, yet salient regional political territory? Or will other cultural and political identities—such as Indigeneity, nationality, race, or class—blunt the emergence of a regional collective identity? These questions are not new. Back in 2001, Carl Abbott reflected on similar questions.[6] He concluded that two important factors might blunt the formation of political and economic integration and the crystallization of a collective Cascadian identity: nationalism rooted in the nation-state system and other competing place-based attachments as well as uncertain political and economic conditions within shifting landscapes of global capitalism. With regard to the prospect of a collective identity, in the intervening years, we have seen the resurgence of nationalism as well as the rise of ethnic nationalism and the white supremacist, religious ideologies of Cascadia. Additionally, it remains unclear if a truly integrated economic and political region will emerge in the face of issues of state sovereignty and intraregional competition for global capital, particularly as governments struggle to recover from the economic consequences of the global pandemic. Yet regional elites, for now, forge on, as demonstrated by the continued forward momentum of the Cascadia Innovation Corridor and the regional high-speed rail project.

Furthermore, the place-making of Cascadia itself is racialized. Indigenous people do not conceive of "land or place as simply some material object of profound importance to Indigenous cultures (although it is this too); instead it ought to be understood as a field of 'relationships of things to each other.'"[7] As such, Indigenous place-making cannot be assumed to be congruent to non-Indigenous place-making. Non-Indigenous people, particularly those invested in a vision of a socially just Cascadia, need to acknowledge both the destructive history of settler colonialism and the fundamental differences between Indigenous versus Euro-American and Euro-Canadian ontology and epistemology of place and land as well as what this might mean for a genuine politics of recognition, reconciliation, and Indigenous sovereignty.[8]

Scholar of contemporary Indigenous politics Glen Coulthard (Yellowknives Dene) writes, "Although [Indigenous] place-based ethics has been worn by decades of colonial displacement, for many it still serves as the radical imaginary guiding our visions of a just political and economic relationship with non-Indigenous people and communities based on principles of reciprocity and mutual obligation."[9] The stories told in this book, of the Duwamish River Cleanup Coalition, the Portland Harbor Community Coalition, anti-pipeline activists, the Puyallup Tribe, the inhabitants of exurban Oregon, and housing justice advocates in Seattle and Vancouver, all point to promising ways that people, sometimes against the state and sometimes with allies in government, are working to build more just political and economic relationships. The proliferation of democracy movements, urban and rural environmental justice movements, and movements to assert Indigenous sovereignty, and the potential for Indigenous nations and settler allies in and outside the state to co-govern, are hopeful signs that the people of Cascadia can come together to collectively build "the institutions and practices . . . [that] have the unique power to establish binding rules for living together."[10] They represent emerging movements of solidarity that recognize and repoliticize the diverse identities, ideologies, and material interests of the region. If there is a Cascadia, it is only brought to life by the hard work of people, enmeshed in the deeply contradictory yet utopian longings of the green imaginary.

NOTES

1 Allen J. Scott, "City-Regions Reconsidered," *Environment and Planning A: Economy and Space* 51, no. 3 (2019): 554–80.
2 Scott, "City-Regions Reconsidered."
3 Patrick J. Smith, "Transborder Cascadia: Opportunities and Obstacles," *Journal of Borderlands Studies* 19, no. 1 (2004): 99–121.
4 In many regards, the split between left bioregionalism and eco-capitalism mirrors the emerging debates between degrowth and ecomodernism. The former argues that the crises stemming from capitalist economic growth necessitates a return to a material economy analogous to conditions in the past when economic growth and commodification did not reign supreme. The latter argues that states can direct capitalism to decouple economic growth from increasing material resource use through eco-technological substitutions. For the ecomodernist perspective, see "An Ecomodernist Manifesto," accessed March 26, 2020, www.ecomodernism.org; and Leigh Phillips, "Planning the Earth System: A Call for a Global Democracy," *Breakthrough Journal*, no. 11 (Summer 2019), https://thebreakthrough.org/journal/no-11-summer-2019/planning-the-earth-system. For the degrowth perspective, see Giorgos Kallis, "In Defense of Degrowth," *Ecological Economics* 70, no. 5 (2011): 873–80.
5 For a good overview of this history, see Nils Gilman, "The Coming Avocado Politics," *Breakthrough Journal*, no. 12 (Winter 2020), https://thebreakthrough.org/journal/no-12-winter-2020/avocado-politics.
6 Carl Abbott, *Greater Portland: Urban Life and Landscape in the Pacific Northwest* (Philadelphia: University of Pennsylvania Press, 2001).
7 Glen Coulthard, "Place against Empire: Understanding Indigenous Anti-colonialism," *Affinities: A Journal of Radical Theory, Culture, and Action* 4, no. 2 (2010): 79.
8 Coulthard, "Place against Empire," 80.
9 Coulthard, "Place against Empire," 80.
10 Jedediah Britton-Purdy, "The World We've Built," *Dissent*, July 3, 2018, www.dissentmagazine.org/online_articles/world-we-built-sovereign-nature-infrastructure-leviathan.

CONTRIBUTORS

ALIDA CANTOR is an assistant professor of geography at Portland State University. She is a political ecologist with a PhD in geography from Clark University and postdoctoral training in water law and policy from University of California, Berkeley. Her research examines resource governance in the western United States, including water governance, law, political ecology, and environmental justice of food, water, and energy systems. She teaches courses on water and resource management, geographies of food, and environment and society.

ERIN GOODLING is a community-based researcher, drawing on urban studies, environmental justice (EJ) studies, critical race studies, and urban political ecology to investigate the politics of homelessness and social and EJ movements. She is a postdoctoral researcher in the Department of Geography at the University of Oregon and the research director at the Western Regional Advocacy Project, a coalition of homeless-led activist groups (https://wraphome.org). She has a PhD in urban studies from Portland State University.

NIK JANOS is an associate professor of sociology at California State University, Chico. He received his PhD in sociology from the University of California, Santa Cruz in 2012. His work combines geography, environmental studies, and history and is exemplified by his work on the dimensions of territory, industrial ecology, and environmental justice, "Urbanising Territory: The Contradictions of Eco-cityism at the Industrial Margins, Duwamish River, Seattle," published in *Urban Studies* (2020). He teaches classes on the environment, globalization, theory, and public sociology and in his spare time ponders the prospects of deep decarbonization. http://nikjanos.org.

CORINA MCKENDRY is an associate professor of political science and the director of the Environmental Studies program at Colorado College. She earned her PhD in politics from the University of California, Santa Cruz in 2011. Her teaching and research focus on questions of climate justice, subnational climate policy, and urban environmental governance. She is the author of *Greening Post-Industrial Cities: Growth, Equity, and Environmental Governance* (2018).

DANICA MILLER is an associate professor of American Indian studies in the School of Interdisciplinary Arts and Sciences at the University of Washington Tacoma and a citizen of the Puyallup Tribe of Indians. She grew up on one of the last continuous allotments of the Puyallup reservation in a highly political Puyallup family, which is reflected in her research and teaching on Puyallup history and Indigenous sovereignty.

MARK PURCELL is a professor in the Department of Urban Design & Planning at the University of Washington, where he studies cities, political theory, and democracy. He is the author of *Recapturing Democracy* (2008); *The Down-Deep Delight of Democracy* (2013); and numerous articles in journals, including *International Journal of Urban and Regional Research, Urban Geography, Antipode, Urban Studies*, and *Planning Theory*. https://home.foreveroverhead.cloud.

JENNIFER L. RICE is an associate professor in the Department of Geography at the University of Georgia. Her research centers on climate justice, urban governance, scholar-activism, and feminist and anti-racist praxis and pedagogy. Her longest research project in Seattle, Washington, has focused on carbon governance, climate gentrification, housing justice, and radical democracy.

ALEXANDER REID ROSS is a lecturer at Portland State University and holds fellowships at the Centre for Analysis of the Radical Right and Political Research Associates. He received his PhD from the Earth, Environment, and Society program at Portland State University on adaptive resource management in hydrosocial territories in 2020 and is currently coauthoring a book about climate change and urban water resources.

JEFFREY C. SANDERS is an associate professor in the Department of History, Washington State University. He is the author of *Seattle and the Roots of Urban Sustainability: Inventing Ecotopia* (2010) and *Razing Kids: Youth, Environment, and the Postwar American West* (2020).

COLL THRUSH is a professor of history and faculty associate in critical Indigenous studies at the University of British Columbia in Vancouver. He is the author of *Native Seattle: Histories from the Crossing-Over Place* (2007); coeditor of *Phantom Past, Indigenous Presence: Native Ghosts in North American Culture and History* (2011); and author of *Indigenous London: Native Travelers at the Heart of Empire* (2016). His current project, *Wrecked: Ecologies of Failure in the Graveyard of the Pacific*, is a cultural and environmental history of shipwrecks, settler colonialism, and Indigenous survivance on the Northwest coast, a region known as the Graveyard of the Pacific.

GIUSEPPE TOLFO was born and raised in Vancouver, on the ancestral and traditional territories of the Musqueam Indian Band, Tsleil-Waututh Nation, and Squamish Nation. He is a practicing planner working at Urban Strategies in Toronto. In 2019, Giuseppe completed a master of arts in urban planning at the University of Waterloo, where he researched public discourses surrounding neighborhood change and gentrification. He remains interested in and continues to write about how the construction of narrative works to reinforce difference and provide justification for exclusion.

THAÏSA WAY, PhD, Cornell University, FASLA, FAAR, is director of the Garden and Landscape program at Dumbarton Oaks, Harvard University, and professor of landscape architecture, College of Built Environments, University of Washington. She researches, writes, and teaches landscape history with a focus on issues of gender, race, and difference in the formation of landscape architecture as a profession and practice. She is the author of *Unbounded Practice: Women and Landscape Architecture in the Early Twentieth Century Design* (2009, 2013), *From Modernism to Urban Ecological Design: The Landscape Architecture of Richard Haag* (2015), and *GGN: Landscapes 1999–2018* (2018), among other publications.

KEN P. YOCOM is an associate professor and chair of the Department of Landscape Architecture at the University of Washington. Trained as an ecologist and landscape architect, he examines the relationships between landscape processes and development practices in the built environments. His research and professional projects have taken him to many parts of the world, but he is most grounded when his feet are firmly planted on the shores of the Salish Sea.

INDEX

C

U

X

Y